T0290353

Private Fire

Private Fire

Robert Francis's Ecopoetry and Prose

Matthew James Babcock

UNIVERSITY OF DELAWARE PRESS
Newark

Published by University of Delaware Press
Co-published with The Rowman & Littlefield Publishing Group, Inc.
4501 Forbes Boulevard, Suite 200, Lanham, Maryland 20706
www.rlpgbooks.com

Estover Road, Plymouth PL6 7PY, United Kingdom

British Library Cataloguing in Publication Information Available

Library of Congress Cataloging-in-Publication Data

Library of Congress Cataloguing-in-Publication Data on file under LC#2010011942
ISBN: 978-1-61149-022-0 (cl. : alk. paper)
eISBN: 978-1-61149-023-7

♾™ The paper used in this publication meets the minimum requirements of American National Standard for Information Sciences—Permanence of Paper for Printed Library Materials, ANSI/NISO Z39.48-1992.

Printed in the United States of America

for my tribe

and those next view, who dwell content in fire
—Dante

Contents

Preface

MY FASCINATION WITH THE LIFE AND writing of Robert Francis has spanned two decades and peaked significantly during two calendar years: 1987 and 2007. During my senior year at Jerome High School in Jerome, Idaho, my mother, a former junior-high school English teacher, presented me with a used copy of the second edition of Stuart Friebert and David Young's *Longman Anthology of Contemporary American Poetry 1950–1980*, which she had purchased for one dollar at the local Idaho Youth Ranch. In it, I read some of Francis's poetry for the first time.

Friebert and Young present an eclectic but representative selection of Francis gems in their volume, including what I consider the *ne plus ultra* of experimental environmentalist lyrics: "Silent Poem." This anthology's black-and-white photo of Francis, taken by Stephen Friebert, presents an iconic and paradoxical view of the author outside his home, Fort Juniper. But the sap of paradox flowed through Francis's veins, as anyone who reads his work discovers. His compositions revel in the thorny paradoxes that meld human and non-human biospheres. In his poem "Paradox," for example, the speaker notes the "teasing paradox" of the wild New England raspberry and how it "hides itself even / while publicizing itself," "half- / concealed in leaves." And the lucid coda: "the same paradox / that I myself / (forgive me) am" (2–3, 5–6, 9–12).

In one sense, the Friebert photograph captures the comic side of Francis: Yankee yokel; simpering scarecrow in baggy second-hand clothes; a fence-post embodiment of St. Francis of Assisi's clown assistant, Brother Juniper. At the same time, the picture conveys the image of a man devoted to ideals of simplicity, economy, hardiness, and low-impact interconnectedness with the natural world. If anything, this candid snapshot locates Francis's "Thoreau nouveau" ethos of self-reliance "outside the twentieth century," as he labels his credo in his unpublished manuscript *Traveling in Concord*.

What I didn't know in 1987 was that I had discovered Francis in the year when he died. A month short of his eighty-sixth birthday, he passed away in a Northampton hospital from complications related to a fall at home. Though I encountered the writing too late to meet the man, I continue to investigate the paradox of my interest in his contribution to environmental writing: Why would poetry and prose rooted in the sylvan landscapes of Massachusetts mean anything to someone from the semi-arid regions of eastern Idaho's Snake River flood plain?

In an attempt to answer this question, I traveled to Amherst in May 2007 to experience Fort Juniper. Outside the DuBois Library on the campus of the University of Massachusetts, I met Henry Lyman, Francis's executor. Henry drove me to Market Hill Road for a tour of the modest mecca and grounds, which, since Francis's death, have served on an informal basis as a secluded retreat for writers. Having read about Francis's home, I was unprepared for my reaction on arrival. Rather than peaceful, I felt edgy, nervous—almost unworthy—as I strolled around the spartan white clapboard house, listened to Henry's anecdotes, and snapped a hasty batch of badly focused digital photographs. After a brief stay (during which I didn't sit down once, though Henry brought out chairs), I found myself requesting that we terminate the visit, as if I had pressing business. Before leaving, however, I asked Henry to read Francis's "Come Out into the Sun" from Fort Juniper's granite front step. Despite the irreverent howl of traffic on Market Hill Road (now a clogged speedway compared to the dirt walking route Francis describes in earlier prose), Henry performed a splendid woodland reading, and I felt temporarily transported in the "deep diathermy of high noon" as an "acolyte of the illumined air" (12, 19). Each night during my stay in Amherst, I returned to Market Hill Road to collect more footage and to take in the spring sights and sounds around Fort Juniper. A palpable aura of solitude enveloped the nearly seventy-year-old one-man citadel built of "hurricane pine" from the storm of 1938. I couldn't stay, and I couldn't tear myself away.

The culmination of my obscure odyssey became this project, the most comprehensive "green" reading of Francis's poetry and prose to date, which is not to say it should be the last. My hope is that my contribution (so incomplete in its "comprehensiveness") will spark further inquiries, books, and articles about Francis's life and his unique gift to twentieth-century American literature. The trove of writing, artifacts, photos, and personal records that remains unexamined, uncollected, and unpublished suggests that there is ample material for additional scholarly activity.

In arranging my material, I sought two objectives: to be accessible to academic audiences as well as general readers, and to emulate—as far as any analytical endeavor *can* emulate a poem—Francis's injunction that a poem should direct its own making, that poets should let poems grow and evolve like wildflowers and woodchucks, that they should be coaxed and watched by authors, but mostly left alone to reach a state of fruition, according to their naturally determined time frames. Not surprisingly, this evolutionary "waiting and watching" (and daily writing) method produced nine loosely related chapters on a smattering of topics.

This topically arranged study, which utilizes biography and literary theory as a set of interpretive lenses, seeks to classify Francis as an unjustly marginalized twentieth-century ecopoet who deserves to be more widely read and studied, especially in the face of such dire environmental and economic straits in twenty-first century America and other countries around the globe. To understand, appreciate, and classify Francis accurately, we must consider the complex interrelationship between the *types* of works he produced and the *way* that he lived and viewed the world. For this reason I offer an examination of an assortment of Francis's published and unpublished works (autobiography, traditional and experimental poetry, fiction, essays, criticism, journals, and correspondence) as seen through an assembly of his distinctly multifarious and sometimes unstable views concerning the influence of his poetic predecessors, sexuality, environmentalism, conservation, spirituality, politics, and pacifism. Only by considering the variety of literary texts that Francis produced in conjunction with the array of world views and living practices that he maintained can we begin to see why he should be regarded as an important twentieth-century American ecopoet.

In chapter 1, my introduction, I delve into the details of Francis's life, particularly those that formed him as a writer. This biographical apparatus is intended to assist readers in exploring his work firsthand, given his relative obscurity. No biographical treatment, however, should be taken as a substitute for Francis's autobiography, *The Trouble with Francis*, a book that, according to Fran Quinn (Francis's co-executor), was nearly optioned as a Signet paperback but was inexplicably cut from Signet's list (Quinn, telephone interview with author, December 8, 2008). Lucid, engaging, witty, candid, and as close to truly objective as any autobiography can get, *The Trouble with Francis* may indeed be Francis's greatest work. While some will disagree, I feel that it deserves to be included alongside similar but more visible books, such as *Walden* (Thoreau), *Studies in the Sierra* (John Muir), *Sand County Almanac* (Aldo Leopold), *Pilgrim at Tinker Creek* (Annie Dillard), *Long-Legged House* (Wendell Berry), *Desert Solitaire* (Edward Abbey), *Refuge* (Terry Tempest Williams), and other "ecobiographies," as Cecilia Konchar Farr has called works of this nature. After spotlighting milestones and summits in Francis's life, I revisit the scholarly assessments of mainstream readers and literary presses, particularly those that contributed partly to Francis's obscurity: beliefs that he was merely a baseball poet, a Robert Frost copycat, a children's author, or a gay nature lyricist. After tracking these few historical niches, I draw upon an eclectic array of ecocritical literary theories and reading strategies in order to suggest an additional category for Francis in light of his diverse environmentalist output.

Chapter 2 incorporates two sections, one on Dickinson, the other on Frost. The first section—"[T]hat this is 'Amherst'": Emily Dickinson and the Phenomenology of Place," traces Dickinson's influence on Francis in terms of poetics, lifestyle, and outlook. In general, Francis's stylistic inheritance from Dickinson and his lifelong outpouring of poems, essays, articles, and autobiographical musings that center on her poetry and life combine to cast Uncle Emily as Bob's master mistress. In order to pursue this connection, I expand on several now-canonized theoretical texts on influence (Harold Bloom's *Anxiety of Influence* and Sandra M. Gilbert and Susan Gubar's *Anxiety of Authorship*) in order to develop the notion of "eco-influence," or ways that geographical *places* shared by authors influence them and their writing. While Francis welcomed his association with Dickinson, he appears initially to have sought out Frost's advice and mentorship, but then drifted from the elder poet if not rejected the influence of Frost's gift outright. In "After Apple Peeling," I sidestep the critical history of superficial comparisons between the two poets and examine an episode that changed both Roberts forever: the intriguing "Apple Peeler" conflict.

Chapter 3, rather than focusing on genre, applies gender ideology and sexual identity (as they pertain to the writer's natural environment) to Francis's prose, poetry, and experimental prose poetry. In "The Gender of Genre," Eve Kosofsky Sedgwick and Daniel Spencer's *Gay and Gaia* assist me in reading gender as a socio-environmental event in Francis's neo-Thoreauvian prose rhapsody, *Gusto, Thy Name was Mrs. Hopkins.* My second section, "The Ecology of Eros," examines Francis's prose-poetic collection of homoerotic nature writing, *A Certain Distance.* In the conclusion of this chapter, "Male Venus, Eco Homo," I piece together several key prose selections from Francis's journals and autobiography that trace the narrative of his life as a gay artist writing in rural New England in the modern and postmodern periods of twentieth-century American literature.

Chapter 4 treats prose—fiction and non-fiction, published and unpublished. The first segment, "Flight," centers on Francis's only published novel, *We Fly Away*, as an example of what Karla Armbruster has called the "bioregional narrative." In an effort to bring awareness to over 160 meditative prose pieces that still remain uncollected, "Evolution" touches on the wealth of essays that Francis wrote for *The Christian Science Monitor* and *Forum* from the post-Depression years through World War II.

Chapter 5 joins religion and politics. First, responding to an assortment of Francis's most devotional and skeptical poetry, I scrutinize his sometimes inscrutable and always heterodox mixture of religious views and philosophies on the spiritual dimensions of nature. Next, I employ Jonathan Bate's assertion that green readings cannot separate "ecopoetics from ecopolitics" to investigate some of Francis's most caustic nationalistic satires and anti-war poetry, the product of his days as a conscientious objector during World War II and Vietnam.

Chapter 6, "Economy, Place, and Space," showcases poems that probe nature's capacity for conservation but that also display Francis's hallmark aesthetic: brevity. One feature of Francis's conservationist poetics was his regard for geographical places as precious resources. Many of these poems emerge as examples of what ecocritics have called "place-making" texts, works that meditate on the generation of relationships between humans and the places they inhabit. As a supplement to Francis's "biosthetic" approach to life and art, his "aesceticism" becomes a method of creating art, with nature as guide, that increases the artist's productivity by reducing wasteful consumption of time, resources, and language. In short, I demonstrate how, for Francis, "poetry" and "poverty" were only one letter apart.

Chapter 7 targets Francis's foray into a mixture of experimental poetics with environmentalism, specifically four sub-genres he labeled "word-count" poetry, "fragmented surface" poetry, "mono-rhyme," and "silent poetry."

Chapter 8 resurrects Francis's one and only long narrative poem, *Valhalla*. A multi-leveled view of literary "environmental apocalypticism" restores this laudable but forgotten narrative poem to a generational line of prophetic works that span from ancient times to the present eco-catastrophic moment.

My conclusions in chapter 9 are retrospective, introspective, and projective. Partly for the purpose of filling in some gaps, I trace Francis's connection to better known writers of his period—among them, Marianne Moore, Donald Hall, James Merrill, Robert Bly, and James Dickey—in order to situate him historically and to contextualize his contribution to twentieth-century literature. I also indulge in some personal reflections on this book's quest-like narrative and the corrective impact that Francis's conservationist aesthetics has had on my experiences as a writer, educator, and global citizen.

Acknowledgments

THANKING PEOPLE IS DANGEROUS BECAUSE someone always gets left out. Numerous individuals, some known but most anonymous, deserve more recognition than I can give them here. Quite simply, without them, this monumental task would not have been possible. At Indiana University of Pennsylvania, first and foremost, Dr. James M. Cahalan, for guidance, direction, and much-needed criticism. Also, Drs. Karen Dandurand and Lingyan Yang, for valuable feedback and coaching. Thanks, too, to the assiduous and affable staff at IUP's Stapleton Library.

At Brigham Young University-Idaho, a special thanks to the College of Language and Letters and the English Department for priceless resources both temporal and material. Their cheerleading in the hallways, department retreats, and moral support provided balm to the walking-wounded soul. Specifically, I wish to thank Deans Rodney Keller and John Ivers, and Department Chairs Kip Hartvigsen, Karen Holt, and Kendall Grant, all of whom saw me through this. For the number of requests I sent them, the media services and interlibrary loan staffs at BYU-Idaho's McKay Library deserve immortality. Two research assistants—Trevor Mason and Karilyn Turner—helped me jumpstart this project in its early phases. And later, Tyler Haynes, Ian Weaver, and Rebecca Jolley, my teaching assistants, helped shoulder the burden of my duties so I could drift where I needed to.

For the walk through time they gave me, my warm appreciation also goes to the special collection personnel in the Bird Library at Syracuse University and the DuBois Library at The University of Massachusetts-Amherst. Thanks to the staff at Dartmouth's Rauner Special Collections Library, too, who sent photocopies of rare materials. Boundless gratitude goes to Tevis Kimball and Kate Boyle at Amherst's Jones Library. The three days they "broke all the rules" gave this project the maverick spirit it needed. For me, it was a once-in-a-lifetime thrill to sit where Francis sat, to write about his life for a few days where he spent years writing about it. I am grateful to Fran Quinn, Francis's co-executor, for phone conversations and priceless anecdotes. Endless felicitations and a lifetime of appreciation go to Henry Lyman for opening his home, personal collections, and Fort Juniper. Undoubtedly, without Henry's selfless contributions, this project could not have been completed. Above all, to my wife, Missy, and our children—Shayla, Tate, Avery, Ceceli, and Crane. I look forward to getting to know you again. Without you, the world would have no poetry.

Finally, to Bob Francis of Amherst, Massachusetts. I didn't get to thank you in person, so let me thank you now. May we meet on a country road somewhere in another life and sit down for the long talk we never had.

Acknowledgment is hereby made to the following institutions and individuals for permission for use of the following:

Reprinted by permission of the publishers and the Trustees of Amherst College from *The Poems of Emily Dickinson: Variorum Edition,* Ralph W. Franklin, ed., Cambirdge Mass.: The Belknap Press of Harvard University Press, Copyright © 1998 by the President and Fellows of Harvard College. Copyright © 1951, 1955, 1979, 1983 by the President and Fellows of Harvard College.

Reprinted by permission of Harvard University Press from *The Environmental Imagination: Thoreau, Nature Writing, and the Formation of American Culture* by Lawrence Buell, pp. 63, 73, 227, 251, 281, 283, 285, 299, 301, 305, Cambridge, Massachusetts: The Belknap Press of Harvard University Press, Copyright © 1995 by the President and Fellows of Harvard College.

The University of Massachusetts Press for excerpts and works from *Robert Francis: Collected Poems 1936–1976* (1976); *Frost: A Time to Talk* (1972); *The Trouble with Francis* (1971); *The Satirical Rogue on Poetry* (1968).

The University of Michigan Press for excerpts from Robert Francis's *Pot Shots at Poetry* (1980).

The University of Wesleyan Press for excerpts from "Come Out into the Sun," "Hallelujah: A Sestina," "Three Darks Come Down Together," "Apple Peeler," "Floriut," "The Rock Climbers," "Farm Boy after Summer," "The Base Stealer," "Swimmer," "Waxwings," "Epitaph," "The Orb Weaver," "The Disengaging Eagle," "Blue Jay," and "Monadnock" from Robert Francis's *The Orb Weaver* © 1960, Reprinted by permission of Wesleyan University Press.

The YGS Group and the Associated Press.

The estate of Fran Quinn.

The estate of Henry Lyman.

To the Pilgrim Press for excerpts from Daniel Spencer's *Gay and Gaia,* Copyright © 1996.

To Elinor Cubbage for excerpts from "Robert Francis: A Critical Biography," Copyright © 1975 Elinor Phillips-Cubbage.

To David Graham for excerpts from "Millimeters Not Miles: The Excellence of Robert Francis, originally published in *Painted Bride Quarterly* 35.1 (1988), Copyright © 1988 David Graham.

Oxford University Press for excerpts from John Gatta's *Making Nature Sacred: Literature, Religion, and Environment in America from the Puritans to the Present*, Copyright © 2004, Oxford University Press.

The artwork of David Fichter.

To Lance Larsen for excerpts from *In All Their Animal Brilliance,* Copyright © 2005, Tampa University Press.

To Cornelia Veenendaal and the former Rowan Tree Press for excerpts from *Travelling in Amherst: A Poet's Journal 1930–1950,* Copyright © 1986, Rowan Tree.

To Stephen Friebert for photograph in the *Longman Anthology of Contemporary American Poetry: 1950–1980,* 2nd Edition, Copyright © 1983, Longman.

To Routledge for excerpts from *De-centering Sexualities: Politics and Representation Beyond the Metropolis,* Phillips, Watt, and Shuttleton, eds., Copyright © 2000.

The Jones Library archives.

The special collections in the Bird Library, Syracuse University, New York.

The special collections in the DuBois Library, University of Massachusetts-Amherst.

To Leonard Lizak for excerpts from "Robert Francis: A Trinity of Values, Nature, Leisure, Solitude," Copyright © 1966, Leonard Lizak.

The Office of Amherst Parks and Commons.

Robert Bly, for copy of "Visiting Emily Dickinson's Grave with Robert Francis."

Private Fire

1
Introduction

PRIVATE FIRE ANSWERS THE CALLS for greater attention to be given to Robert Francis and carves out a foothold for him in the study of American ecopoetics.[1] Across his life's long arc, Francis's primary subject remained the same: the earth and its relationship to natural, cosmic, and atmospheric phenomena. From his first days as a writer, he saw that his call as an artist was toward the preservation of the ground on which he walked, the sunlight and streams in which he bathed, and the air he breathed. Throughout his career, though he experimented with other subjects, he gravitated back toward probing the nature of the human-earth relationship. In a 1936 poem entitled "Clouds," he transforms a sweeping vanguard of New England clouds into a contemplative vehicle of transportation that allows humans to both escape and remain in partnership with a troubled biosphere: "Moving with them we move beyond all ills—/ Far from the ailing earth, yet not too far" (11–12). In using the term "ecopoet" to describe Francis, I do not mean to suggest that only his poetry defines him, nor that an "ecopoet" refers to only one kind of writer. Much like the writing that he constantly revised, Francis evolved, traveled through various phases, and demonstrated that to be an ecopoet meant to be not one thing but many.

Today, depending on whom you approach, if you ask, "Who was Robert Francis?" you receive either a variety of clipped responses or bewildered looks. The question is not easy to answer, but it is worth the pursuit. A native of Upland, Pennsylvania, born literally at the turn of the twentieth century, Francis was raised by his mother, Ida May, and his father, Ebenezer, a Baptist minister, in Camden and Toms River, New Jersey; Greenport, New York; and Dorchester, West Medford, and finally Amherst, where at the age of ten he "became a New Englander" (*Trouble with Francis*, 148). His poem "New England Mind" describes a lifelong intermingling of northeastern soul and landscape. "My mind matches this understated land," the speaker exults, "Outdoors the penciled tree, the wind-carved drift" (1–2). Though energized and enlivened by local geophysical features and climate, the speaker finds that the woods, fields, and streams outside his back door act as a metaphysical bridge to more global, even universal, imaginative experience. A hybrid local and global worldview, or "glocal" outlook, as James Cahalan has referred to it in his study of hometown authors, leads Francis's speaker to a meditative parapet of epiphany (Cahalan, 250). "My outer world and inner world make a pair," he concludes, wondering if he would "be New England anywhere" (13, 16). Marianne Moore, writing to the University of Massachusetts Press, recognized the ripe regionalism in Francis's blood. "He belongs with Emerson and Thoreau," Moore observed. "I think him one of our indispensable, truly typical New England authors. . . . Mr. Francis is New England through and through" (letter to Leone Barron, December 31, 1966). Though he traveled a migratory route to settle there, the northeast's geography would claim Francis as a native.

By his own account, Francis's youth and adolescence brought a mixture of blissful transcendence and traumatic shock and alienation. His autobiography relates accounts of a sheltered home life, rapturous swims in the sea, rowing trips with his father, a sighting of the 1910 Halley's Comet, and Wordsworthian outdoor rambles in whatever natural settings he encountered. At the same time, he remembers himself as "a scared cat and a cry baby," a musical and artistic prodigy, a boy inclined to play with dolls and to eschew the rough-and-tumble company of classmates, someone who by the sixth grade recognized the existence of a social "elite to which [he] did not belong" (*Trouble with Francis*, 144, 154). In the era before television and radio, language and literature close at hand stimulated his creative impulses. "Night after night I died happily without being conscious of death," he recalls, describing his mother's ritual of reading aloud to him as he drifted to sleep (*Trouble with Francis*, 154). In his late thirties, he reflected on his gradual development in "Biography," a trio of clipped quatrains in *Valhalla and Other Poems*: "Speak the truth / And say I am slow, / Slow to outgrow / A backward youth. / Slow to see, / Slow to believe, / Slow to achieve, / Slow to be. / Yet being slow / Has recompense: / The present tense. / Say that I grow." Remarkably, Francis was so slow to develop in some areas that, as a child, he never learned to ride a bicycle—a task he finally mastered at the age of forty-one! "I regard slowness as a principal key in my life," he would later observe with philosophical satisfaction (*Trouble with Francis*, 47). Highly sensitive and shy as a youngster and young man, he would live to tap that reservoir of sensitivity and with a tenacious resolve face life's uphill slope and generate a timeless body of literature.

In terms of faith, he was a believer whose belief evolved. Initially devoted to his father's spiritual orientation, at least outwardly, as a young man he subjected himself to a regimen of almost Puritanical physical, intellectual, and spiritual improvement rituals he called his "Rules & Regulations." This procrustean gauntlet of requirements demanded over twenty daily duties, including cold baths, memory improvement, attendance at Sunday and Thursday church services, deep breathing and stretching exercises, strict bedtime observance, and spotless personal hygiene. Records from his teenage years reveal private obsessions with the Book of Job and library self-improvement books, such as Webster Edgerly's *Ralstonism* and Frank C. Haddock's *Power of Will*, the latter of which his mother blacklisted and took from him, as if it contained something satanic that was destroying his soul. Reminiscent of Stephen Dedalus's mortification-of-the-flesh episodes in *A Portrait of the Artist as a Young Man*, a spate of journal entries details a tortured series of psychologically self-flagellating scenes in which the young Francis berates himself for excessive gluttony and indolence.

Still, wedged within this austere loyalty to transcendence through unflinching asceticism, the budding ecopoet registered prophetic signs of his eventual efflorescence into the world of art and the art of the world. Item ten in his "PHYSICAL" rules and regulations section calls for a "special hour outdoors," and item three in the "INTELLECTUAL" category requires "three new words mastered daily" (*Trouble with Francis*, 164). Reflecting in his seventies on his religious upbringing, he observed, "How curiously misguided I was in much of this," having superseded a mostly "Roman" religion "with no true inwardness . . . and certainly no joy," a willed worshipfulness that was "all duty and observance." Having explored a lifelong range of self-denials and self-indulgences, he wrote of his youth, "Surely some of my rules were more beneficial in the breach than in the observance" (*Trouble with Francis*, 166–67). His absorbing spiritual odyssey spans from the time, as a teenager, that he wrote "Do the hard thing" on his baptismal certificate to his formulation of the panentheistic "Religion of One's Own" in *Traveling in Concord* and, at eighty-three years of age, the stubborn shoehorn skepticism in *The Trouble with God*. Whatever the reason, his evolution from loyalty to his father's lockstep Baptist theology toward embracing a more nature-oriented "ecospirituality," as John Gatta terms it, surfaces and saturates Francis's published and unpublished works (Gatta, 7).

The candor with which he delivers the story of his altered convictions preserves precious ironies and inspiring personal dramas. He purposely does not swerve from emphasizing that the typewriter that produced his first published book of poems, *Stand with Me Here* in 1936, was a gift from his father's congregation in Lynn, Massachusetts. His account of how he remained devoted to his father while turning from his father's devotion casts him as a man of not only fervent artistic drive but uncommon courage, intellectual power, and emotional attachment. In the ecstatic "Hallelujah: A Sestina," he exults with the energy of a psalmist, "Could I ever praise / My father half enough for being a father / Who let me be myself? Sing Hallelujah" (25–27). Even in his eighties, in his iconoclastic *The Trouble with God*, he writes that religion was never the cause of "strained relations of any kind" between himself and his father.

"In this statement of unbelief," he declares, "I believe I am honoring my father. The word 'truth' was often on his lips. What better would he have wanted in his son than that he too would be a truth seeker?" (113–14). As a marginalized American writer, Francis clung to a raw, organic faith that fueled his struggle up the Jacob's ladder of his life—then he disassembled that ladder and with the fragments erected an array of metaphysical bridges, spiritual sylvan shelters, aesthetic country turnstiles, and garrets in the air. "Nature is my mother and I am wholly her child," he was able to declare near the end of his life. "What I have found is that the world I live in provides human beings with infinite blessings, infinite tortures, and everything in between" (*Trouble with Francis*, 234–35).

He attended high school, college, and graduate school, but the world provided his education. Despite serving as editor-in-chief of his school magazine and delivering the first four-year straight-A record that Medford High School had seen, he confessed that, even as valedictorian of the class of 1919, he possessed not "the tiniest spark of brilliance" (*Trouble with Francis*, 174). Twice he attended Harvard, first as an undergraduate, then as a graduate student. But he repudiated the distinction. "Harvard's great gift to me was a negative one," he declares. "It didn't make me a Harvard man. I was never captured or committed" (*Trouble with Francis*, 183–84). A year of teaching abroad at the American University's prep school in Beirut and a year at home to attend to his mother during her passing divided his bachelor's degree in history and master's in education, in 1923 and 1926 respectively. Despite awards from peers, publications in *The Nation*, and Robert Hillyer's praise for his writing, Francis depicts Harvard as a hall of despair. "I did not fulfill myself in any way," he writes. "Nothing about college gave me pleasure or satisfaction." His senior year he remembers as the "loneliest year of [his] whole life" but one that, like a crucible of experience, ignited a desire to launch his literary arc: "Out of my loneliness, I wrote a few little poems" (*Trouble with Francis*, 176, 183). Scattered journal entries plot his slow pilgrimage away from institutionalized education toward individualistic geocentric enlightenment. In one entry (May 2, 1934), he recalls how he "celebrated May Day by taking the sun and planting herb seeds" and takes care to contrast local red maple blossoms with the "pale green gold of the rock maple blossoms," after which he juxtaposes this seasonal delight with a recent lecture on Blake that he attended at Mount Holyoke, an experience that "did not prove to be particularly spring-like." He exclaims, "Oh, these lectures! One is forced to ask the embarrassing question: Does this lecture offer more than what could be gained with less time and effort and no expense from a book?" (*Travelling in Amherst*, 33). In another entry, he writes, "At college we enjoyed the vague supposition that we could have everything. College, indeed, seems to be an institution for inculcating this very notion. . . . There is no course in choosing, no course in learning to go without" (*Travelling in Amherst*, 41). In all, these vignettes provide a glimpse of the reluctant teacher-in-training and his chrysalis-like emergence as self-reliant student of the world.

His education, like many aspects of his life, wove a pattern of contradiction and metamorphosis. His poetry tracked the chronology of this evolution. Despite his private feelings of antipathy, a decade before he confessed in print that he never felt like a "son" toward his alma mater, he was elected an honorary member of Phi Beta Kappa at Harvard and commissioned to write and read "The Black Hood" (*Trouble with Francis*,184). In 1955, five years previous to Harvard's award, he was elected Phi Beta Kappa poet at Tufts. A masterful formalistic arrangement of eight rhymed sextets whose stately syntax cloaks an audacious and satirical critique of higher education's masque of deceit, "The Black Hood" unravels the university's most recognizable symbol of ceremony and, with playfully sinister sleight-of-semantics, proclaims it fitting headgear for both graduates and grim reapers. The poem's idealist-realist speaker yearns for "[s]omeone uncommitted as fresh air" who "dare[s] to tell . . . [c]lean truth from trick" (21–22). Higher education's duplicitous "hood-hiding" and "hood-winking," the speaker laments, fosters an elaborately lyrical self-deception: "How many times we think that we are thinking / In making believe we do not make believe" (28–29, 34). Ultimately, however, the futility of the marginal writer's hankering for simple truth mimics the university's cult of smoke and mirrors: "How guileless that black hood compared to me" (48).

Liberated from Harvard's gilded prison in the mid-1920s, Francis then spent decades reading, writing, and basically living in Amherst's Jones Library, which he used as "exhaustively as the Bedouin makes use of the camel" (*Trouble with Francis*, 206). Today, the Jones Library, the house that built Francis, maintains priceless archives of his manuscripts and memorabilia. Though his formal education largely contributed to and postponed his more valuable informal and self-generated artistic education, his aversion to educational institutions did not prevent him, at the age of forty-five, from enrolling at Massachusetts State College. Like William Stafford, Francis spent World War II as a conscientious objector, receiving "1AO classification" and working mostly as a headquarters office staff member (because of his typing skills) and teaching enlisted men rudimentary reading and writing comprehension in night school (for which he was promoted to private first-class) (*Trouble with Francis*, 34, 38). His inglorious military service, however, did not stop him from using the GI benefits he earned to pay for two additional university courses in his forties, not in literature, but in botany and geology. Throughout his life, his omnivorous mind and restless soul hungered for the individualistic real-world learning that would make him a writer. The greatest lesson he learned, the one his life story teaches, is that real learning leads from the lure of the lecture hall to the lay of the land.

It is difficult to categorize Francis's "professional life." "Professional" does not describe the wayfarer's vector he cut through the early twentieth century. More accurately, he drifted, subsisted, and kept vigil for his true self. After Harvard, he slipped from conformity to church and school to the gradual experimentation with and ultimate acceptance of a laissez-faire life of disciplined recreation, of Thoreauvian sauntering and Whitmanesque lounging. His words concerning his train ride to Amherst on Labor Day in 1926 portend a Dickinsonian societal withdrawal: "I was a nature observer in a rudimentary way, a sunbather, a man of peace opposed to war, even something of a poet" (*Trouble with Francis*, 189). For six formative years, he reveled in a "second boyhood," living in his father's parsonage—writing, reading, exploring nature firsthand—years he called the "happiest of [his] life." Though he describes this interval as "a little disgraceful," he credits it for bequeathing him a lasting sense of "emotional equilibrium." "Outdoors," he reminisces, borrowing the rogue phraseology of Walt Whitman, "I had the whole landscape to explore—country road, wood road, wood path, pasture, meadow, marsh, stream, hill. I could go wherever and whenever I wished. I strolled, I sauntered, I rambled. I sunbathed and water-bathed. I loafed and invited my soul" (*Trouble with Francis*, 195–96, 197). During this holistic hiatus of mind and body, he pursued the art of the avocational while his foray into employment, according to typical early twentieth-century social expectations, bore a conspicuous lack of personal drive.

Even when he struck out to wend his way, he plotted a clearly unconventional path. In 1932, at the age of thirty-one, he left the parental roof, moved to Amherst proper, and taught violin lessons while serving as an odd-job man in the homes of elderly widows as a way of earning bread and board. These in-house encounters with feminine domestic life led to several thematically related poems and his novel *We Fly Away*. One bumptious landlady, Margaret Sutton Briscoe Hopkins, introduced him to Robert Frost, a moment that drastically altered the trajectory of his career, so much so that he revisited it again and again throughout his writing, and specifically in his prose rhapsody, *Gusto, Thy Name Was Mrs. Hopkins*. He taught English at various schools—Amherst High (an "unlovable mob"), Mount Holyoke, then Massachusetts State College (the future University of Massachusetts-Amherst), and Lake Placid's Northwood School—none of which provided him with careers because, despite his credentials, he felt ill-suited for the positions (*Trouble with Francis*, 192). A succinct note to the administration at Belchertown State School captures his sentiments: "Dear Dr. McPherson: I am writing to thank you for giving me work at the school and for letting me go when the experiment was over" (letter to George E. McPherson, January 1, 1932). Amid sporadic teaching jobs, he gravitated toward outdoor labor. After being discharged from the army, he performed short-term government-supervised work in apple orchards and on a chicken farm. Like Thoreau, he was an educated man out of place in the halls of formal education. So for the greater portion of his life, he fled to his academy in the woods and scraped together a pauper's existence by literally living off the land, guest lecturing, giving readings, and publishing poetry and essays in *Forum*, the *Saturday Review, New Yorker,* and *Christian Science Monitor.* By all accounts, he closed his eyes, jumped, and made independent living his lyceum-at-large.

The extent of Francis's poverty is difficult to fathom. The fact that for over fifty years he not only survived but thrived at income levels just above, at, or far below most twentieth-century standards of living invites both admiration and wonder. A line from Goethe, used as an epigraph in *Traveling in Concord*, communicates his philosophy: "In dieser Armut welche Fülle! In diesem Kerker welche Seligheit" ("In this poverty what wealth! In this prison what blessedness!"). Through sheer will, he was able to turn his lack of material things into, as Charles Sides observed, "considerable internal wealth" (i). "A man can live without a job," Francis begins the Thoreau-like economy chapter of his autobiography, "can live without a wife, can live without God, but he can't live without money" (*Trouble with Francis*, 214). Even with inflation rates factored in, the *annual* income amounts he scrupulously recorded are shocking: $489.40 for 1952, $349.80 for 1953 (*Travelling in Amherst*, 85–86). At one point in 1942, the United States draft board refused to believe him when he reported that he and his young boarder, Forrest Sanborn, had been living on five dollars a week. Largely, Francis's commitment to "not earn much but to spend little" proved a rigid but salvific code for living (*Trouble with Francis*, 214, 215, 218). He solved the puzzle of modern life by eliminating the spectral lust for conveniences and amusements. He washed and mended his own clothing. He gathered firewood and food from the woods around his home. He gardened, prepared his own meals, and did not eat fish, fowl, or flesh. He depended on few, if any, electrical appliances, used the library instead of buying books and magazines, and walked instead of driving when he could. Richard Gillman, who edited Francis's published journals, recalls how in 1948 Francis rattled up in a 1931 Chevrolet to meet him at the Amherst Common bus stop (Intro., *Travelling in Amherst*, viii). In 1958, when Francis returned to Fort Juniper from his year-long fellowship in Rome, he found that his telephone service had been cancelled. Instead of restoring the service, he left it unconnected for eleven years until his age and health required him to

renew it (*Trouble with Francis*, 117). Rather than let the twentieth cen-
tury's inflated needs find him wanting, he found that a higher standard
of reduced wants was all he needed.

This New England parsimony pinched his pockets early on. Later,
though, because he refused to raise his needs to the modern world's
hyperbolic levels, it enriched him. The Depression decimated peo-
ple addicted to overblown standards of living, but it hardly touched
Francis. "Thus I spent the winter and spring of 1933," he writes.
"The Depression was at its lowest, but I hardly felt it. How could I
fall when I was already on the floor?" (*Trouble with Francis*, 201). In
March 1937, he notes that he paid no income tax and did not bother
to file a return! (*Travelling in Amherst*, 50). Later, in an interview
with Fran Quinn and a student who was performing an independent
study course on depression-era writers, Francis, with his character-
istic mixture of humor and honesty, remarked that he went through
his "personal depression three years before the country did" (Quinn,
telephone interview with author, December 8, 2008). Throughout his
life, he continued to burn conservative amounts of wood, oil, and coal
to heat his home and cook his meals, and he used only the electricity
he needed. In his master's thesis, Leonard Lizak describes the jer-
ryrigged window box in which Francis would store perishables dur-
ing winter months outside his kitchen window so he could unplug his
refrigerator. Even Francis's refrigerator promoted conservation. It was
a "Little Giant," a compact appliance that, in Francis's words, took up
"no unnecessary space" in his kitchen and consumed "no unnecessary
electricity." In describing the simple but liberating use of his exterior
"coldbox" for food storage during the "cool half of the year," Francis
brims with practicality: "What a satisfaction to put to good use a little
of the vast amount of coldness that nature gives away each year" (*Trou-
ble with Francis*, 216). Lizak notes that during the massive electrical
blackout of November 1965, Francis stood unaffected, independent,
and unencumbered "like a Prometheus at his vantage point upon a rise
to the left of the road, . . . his twenty-five years of preparation . . . sud-
denly rewarded" (29). In the 1960s and 1970s, when his income from
writing began to increase, by keeping his standard of living at a 1930s
gauge, Francis was able regularly to transfer thousands of dollars from
his checking account to his savings. Finance sheets he maintained still
attest to the marvel of how he flourished on nothing but what the
world dropped at his feet.

Ultimately, the latter portion of his life, his unconventional lifestyle,
and his literary output define his legacy. Francis's triumph speaks to the
significance and power of the one-man quest to live in harmony with
and struggle against self and surroundings, the epic modest in scope

but limitless in perspective, a life infused with a love for independence, nature, and writing. Lizak's thesis, "Robert Francis: A Trinity of Values—Nature, Leisure, Solitude," highlights the ecotrinity that Francis espoused as a substitute for the Biblical one with which he was raised. This triad—a spiritual and aesthetic parallel to Dickinson's bee, butterfly, and breeze—provided Francis with his ideal living status, one characterized by a "balance between people and no people, between literary people and plain people." His personal litany flows with all the sublimity of scripture and the emancipating verve of a constitutional declaration:

> By "nature" I meant the whole heavens and earth except for those spots where human concentration was a blight and poison. I meant sun and sky and clouds and hills and rain and rivers and snow and country roads and farms and country people. By "leisure" I meant not the absence of work but time for the work I most wanted to do. I meant the obliteration of the line between work and play. By "solitude" I meant the freedom to choose from hour to hour whether to be with friends or alone. (*Trouble with Francis*, 18)

In an interview with Philip Tetreault and Kathy Sewalk-Karcher, Francis explained how, by inching forward "step by step" and "stage by stage," he met his needs with "available resources and opportunities." The first three items on his explanatory list claim that his move to Amherst, his abandonment of teaching as a career, and his work as a live-in assistant to elderly women gave him a rural lifestyle, leisure time, and financial independence. His final two items mention crucial moves: "d) I gained other kinds of independence when I rented the 'old house by the brook' in 1937. e) With the building of Fort Juniper in 1940 I had at last a completely congenial setting for my life: nature, leisure, privacy, and independence of every sort" (*Pot Shots*, 122). The bare-minimum subsistence he endured and enjoyed in his last two places of residence—excluding the Cowles Lane apartment he occupied briefly in the mid-1970s for health reasons, selected because it provided a view of Dickinson's grave—typified his conservationist approach to life as much as it symbolized his views on art and poetry.

It is hard to imagine Francis's literature without his companion *modus vivendi*. Harder still to imagine would be his life and literature without the land he loved. His move from his old rented millworker's shack, which had provided him no electricity or running water, led to a life of simple luxury at Fort Juniper, which was built with an insurance payment his stepmother gave him after his father died. Francis named his home after hearing Churchill announce the fall of Singapore on the radio. About this time, he adopted the juniper as a "coat of arms" and assumed an ecological identification with the lowly but hardy evergreen:

Somehow I fitted into the ecology. I was part of the unspectacular land-
scape. . . . As a symbol of what might be called unfallability I thought of
the common pasture juniper, the lowest-growing of all our native ever-
greens. How could it fall when it was already close to the ground? Pines,
hemlocks, cedars, spruces would fall if a wind were strong enough, but
a wind could uproot a juniper only by blowing away the soil beneath it.
For years I had been admiring the juniper, after learning to look at it
closely, admiring not only its tough and stubborn character but its beauty
of form and its subtlety of color. (*Trouble with Francis,* 30–31; *Travelling
in Amherst,* 56)

During the almost fifty years that Francis took root and flourished in the
"old house by the brook" and Fort Juniper, his habitations served as ref-
uges, fortresses of solitude, and palaces of realism and enlightenment.

The name Francis chose for his final dwelling-place combines, para-
doxically, militarism and conservation. At times, it provided him protec-
tion from civilization's dark side and, at other times, kept him a prisoner
of the moment. In the mid-1940s, a journal entry defines the ethos of
his one-man soul bunker. "The name reminds me of the reality behind
the name," he notes. "My home is my fort in my fight against obstacles
that besiege me. Like the juniper I have dug in here and am trying to
endure" (*Travelling in Amherst,* 56). While undergoing basic training at
Fort Breckenridge, Kentucky, he nightly summoned the specter of Fort
Juniper as a psychic shelter: "When finally I succeeded in getting under
my blanket without mishap I would draw my head under too and return
in spirit to Fort Juniper. My fort never defended me better" (*Trouble
with Francis,* 38). Years later, however, during an interview with Leon-
ard Lizak, Francis fell silent, stared into the fireplace, and remarked stol-
idly that his home was not a "fool's paradise" (Lizak, 50).

So central was Francis's home and totemic evergreen to his physical,
psychological, and spiritual development that he authored three similarly
themed poems in an effort to meld the histories of his life, home, and
kindred tree. "Fort Juniper," composed during Christmas 1945, contem-
plates the isochronal interrelationships among international, natural, and
artistic struggles when "peace endures" and "war endures, too" (4–5).
The poem takes into account World War II: "the only sound he ever
hears of war," "A flight of bombers" (7–9). Within the phenomenon of
human war, it blends the conflicts in the natural world: "the internecine
pattern of trees," "the trees are warring / Down at the roots" (15–16). In
conjunction with natural and human-made war, the poet includes the
battle of the poet: "the writer's war to make what he says true, / To get it
said, utterly, and in form" (17–19). Elsewhere, in "Juniper," the speaker
declares, "Here is my faith, my vision, my burning bush" (41). And "Sing
a Song of Juniper," a jubilant air which David Leisner rendered as a

contemporary classical guitar composition, celebrates the juniper's "song composed of silence," a song "seldom sung, / Whose needles prick the finger / Whose berries burn the tongue" (2–4). The last lines of "Sing a Song of Juniper" communicate how the natural world reflects and shapes the speaker's introspective outlook: "And gives me outdoor shadows / To haunt my indoor house" (19–20). For the majority of his life, Francis's home served as a contemplative workshop retreat where he challenged and dissolved the metaphysical boundaries that define human and non-human biospheres.

Though camouflaged, his anchorite aesthetics drew the attention of obscure and more famous individuals. However, not everyone under-stood Francis's solitary habitation and his ecosystemic identification with the juniper. Somewhat judgmentally, Elinor Phillips-Cubbage dismisses Francis's monumental earth-centered hermitage in her thesis, "Robert Francis: A Critical Biography," and depicts Francis's lifestyle as a cow-ardly retreat, suggesting that Francis lacked Robert Frost's "juniperian toughness" (Phillips-Cubbage, 143). But it was a curious Robert Frost who cropped up conspicuously—and perhaps enviously—across three decades to monitor how the younger writer was faring in his makeshift Edenic bunker. As it turned out, Frost's constant masquerade of noncha-lant surveillance was not unmerited. In total, Francis produced ten books of poetry, some self-published, some which Frost personally praised, all of which culminated with Francis's *Collected Poems 1936–1976* and the posthumous *Late Fire, Late Snow*. He authored a novel, a memoir on Frost, a prose rhapsody, two collections of satirical essays, voluminous journals, published and unpublished articles on Frost and Dickinson, and his autobiography the year after the University of Massachusetts-Amherst pronounced him Doctor of Humane Letters, *honoris causa*, in 1970. The Juniper Prize for original poetry and fiction, established by the University of Massachusetts Press in Francis's honor, is still awarded today.

Currently, the archives at Syracuse University, the Jones Library, and the University of Massachusetts-Amherst contain recordings, photo-graphs, unpublished essays, manuscripts, correspondence, and journals that could—and should—be re-issued in a multi-volume set of the Fran-cis corpus. During his life, Francis received numerous awards, includ-ing the Shelley Memorial, Brandeis University's Creative Arts Award, an Amy Lowell Traveling Scholarship, and a fellowship in the Academy of American Poets. He also gave readings and seminars at elementary schools, high schools, and universities, and private workshops in Vermont, Kentucky, California, Michigan, and New York's Chautauqua Writer's Workshop, where he forged his friendship with Donald Hall. Despite all the hazards of the twentieth century, which slashed the average male lifespan to forty-five years for those born in 1901, Francis flourished and

nearly reached ninety, an age that Indiana University longevity specialist
James C. Riley confirms as "well above the average life time for people
born the same year" (e-mail to author, December 3, 2004). After Francis
died, trustees and friends scattered his ashes across Emily Dickinson's
grave and other parts of the earth.

So this question arises: How do we situate Francis in the twenty-first
century? To which movements did he belong in the twentieth? In which
literary camp does he linger now? Chronologically, Francis would join
the "cadre of poets" that Eric Haralson, in *Reading the Middle Genera-
tion Anew*, groups as "the most obvious definition of 'middle' . . . since
their seasons of prime productivity fell between the 1940s and the 1970s"
(Haralson, 1). More specifically, Francis belongs to a subset of a "lost
middle generation," since his years of heightened poetic output match the
time frame of those poets Haralson examines—Roethke, Jarrell, Lowell,
Hayden, Bishop, and Berryman—though Francis has yet to garner the
critical attention of his contemporaries.

Francis's status as "lost" and misread stems from several factors. After
1945, and more markedly after 1960, the collective but unwitting myopia
of critics, anthology editorial boards, and teachers gradually constructed
a kind of Franken-Francis monster: one-fifth baseball poet, young adult
author, gay modernist, minor formalist, and Robert Frost imitator. Those
who narrowly called Francis a baseball poet may have reacted to favor-
able critical responses to his 1960 comeback collection, *The Orb Weaver*,
which, in addition to several other sports poems, featured three baseball
poems—"Catch," "The Base Stealer," and "The Pitcher." For over fifteen
years before *The Orb Weaver*, Francis was virtually silenced by the frag-
mented and highly nuanced tastes of an emerging postmodernity—the
"waves of doctrine" that Donald Hall says whisked readers around like
"little literary Vicars of Bray, table-hopping from Andrew Marvell to Pablo
Neruda, from Emily Dickinson to Charles Bukowski" (Hall, "Two Poets,"
125). Thus, a generation of readers after World War II may have mis-
taken Francis's 1960 encore for his 1936 debut. Francis's misclassifica-
tion as a baseball poet may have been due to teachers who saw—and still
see—his simple style and athletic subjects as ideal material for classroom
use.[2] However, Francis himself, ever the existential endurance athlete but
perhaps the least sports-oriented of individuals, clarified that his athletic
poems were veiled meditations on male homoeroticism and metaphors for
the writing process (*Trouble with Francis*, 148).[3] Nevertheless, into the
late twentieth and early twenty-first centuries, educators and publishers
continue to read Francis through a bubble of recycled criticism and to
misrepresent him as solely a baseball writer (see Appendix E).[4]

Francis was certainly not exclusively a sports writer, but it would not
be too far afield to read him as a (post)modern poet of the homoerotic

or as a writer of juvenilia, though either category alone does not suffice. In this regard, mostly overly-enthusiastic anthologists, politically motivated critics, and revisionist culture warriors have shaped this part of his destiny. For example, Allan Wolf's *It's Show Time* includes Francis's "The Base Stealer" in a book devoted to teaching grade-school children about poetry through dramatic performance. Though Francis did author an unpublished manuscript of light verse for children, *Father Gander's Nursery Rhymes*, and published early pieces such as "A Boy's November" in *The Target: A Paper for Boys,* his work does not deserve to be banished to the juvenile sections where some public and university libraries currently house him. Jim Elledge's *Masquerade: Queer Poetry in America to the End of World War II* features Francis as a gay poet writing before the war. Though he selects samples of Francis's more sober realist poetry, Elledge unfairly lops off the latter and most fertile half of Francis's poetic outpouring after the war when, according to David Young, Francis's "best poems were written" (Young 63). Perhaps most bizarre of all, in the company of these absolutist labels, is the inclusion of Francis's "Silent Poem" in *Poetry for Dummies* as an example of how poetry readers can "treat white space as time" (Timpane, 26). In aggregate, this clash of categories tends to cast Francis as, if not a novel non-entity, a fractured collage of personae. Though he attempted at first but never succeeded as a children's author, and even though his contribution to twentieth-century literary homoerotica and gender awareness remains unique and unexplored,[5] such slim classifications belie the totality of his literary vision.

Relegating Francis to the company of so-called minor American formalists only reinforces the mechanical politics of literary marginalization and supports status quo definitions of who is important and who is not. Does Francis rank only as a "minor" writer? If so, to whom? Why? In what way? And does the descriptive tag "minor" always connote inferiority? Is John Clare a minor British Romantic? Might we see a writer who sidestepped the mind-and-body-snatching trends of twentieth-century publishing conglomerates as superior in some ways—in principle and lifestyle, as someone who refused to sell out?

When reviewers and scholars rank Francis lower than other writers, the terms "minor" and "marginal" surface frequently.[6] Still, Francis's more intimate acquaintances, including Robert Frost, have debunked those major/minor pejoratives and observed that Francis's putative "minor" status constituted, if not a distinctive strength, at least not a shameful liability.[7] Francis himself, at different times, assumed various faces in response to those who insisted on padlocking "minor" around his neck on a gilded chain. "I am a poet, minor. Or I try," he sighs in "The Black Hood" (34). More neutral in his early seventies, he muses on his shrinking social category: "I find myself a minority of a minority, perhaps a minority of one" (*Trouble with Francis*, 91). Five years after this auto-biographical observation, now jubilantly defiant, and with his hallmark blend of subterranean satire and frankness, he responded blithely to an invitation to participate with 499 other individuals randomly selected by computer in a University of Southern California doctoral study grandiloquently and awkwardly titled "Curricular Concomitants in the Educational Backgrounds of Men Leaders in Education, Fine Arts, and Literary Fields." Francis responded this way: "I sat down . . . , glad to take half an hour or more to contribute my 1/500th to your doctoral project. . . . But I am not a leader of any sort. . . . It would be no exaggeration to say that my life has become increasingly allergic to computers. Sorry not to be helpful but not sorry to be who and what I am" (Letter to Edward Rutmayer, December 7, 1976). Whether major or minor, he was a master of the *quip pro quo*.

Informed by personal experience, his satirical short essay, "Major," uses the self-reflexive humor of a mock dialogue to deflect the counterproductive scrutiny associated with major/minor classifications. In this sardonic gem, the first-person speaker asks a friend to define "major poet," a task the friend declares "easy" by responding with the dubious statement, "A major poet is any poet of major importance." In reply to the friend's insistence that "quality" rather than "quantity" determines major-poet status, the speaker visibly irritates the friend by replying, "I suppose a fine poet who wrote only brief poems would have to be very fine indeed to be a major poet." At the conclusion, the speaker thoroughly exasperates his friend by asking for means to tell the difference "between a grade-A poet who writes brief poems and a grade-B poet who writes long ones" (*Pot Shots* 154). Whether or not Francis ranks as a major or minor poet, as an underappreciated and largely unknown writer, he resides in what Lawrence Buell calls the "era when ecocritics are still only starting to explore minority canons" (Buell, *Future*, 118). In Francis's case particularly, we might do well to heed Eve Kosofsky Sedgwick's admonition to eschew the "chronic modern crisis of homo/heterosexuality definition" and its "ineffaceable marking" of the categories "minority/majority" (Sedgwick, 11). For my purposes here, the open categories I have assembled around Francis recall crucial factors in his life: the natural world, social and geographical habitation, gender/sexuality, economic status, time period, and nationality.

To compare Robert Francis to Robert Frost is both inevitable and potentially misguided from the start. In some ways, mostly structural and stylistic, they are alike as two tramps; in other ways, as dissimilar as oven bird and oven mitt. Historically, for every critic who pegged Francis as pitcher-poet, queer lyricist, or children's author, four or five faulted him for being faux Frost.[8] At this point in history, there appear to be two schools: those, such as Lawrence Buell, who lump Francis in with a "continuation of talented if subgalactic latter-day poets who are self-consciously Frostian,"[9] and those, such as Donald Hall, who feel Francis "wrote his way past" Frost[10] (Buell, "Frost" Hall, 117; "Two Poets," 119). Undoubtedly, no other American poet influenced Francis more than Robert Frost during the thirty years they knew each other. I hesitate, however, to agree with the benign term of "friend" that Edward Ingebretsen uses to describe Francis in *The Robert Frost Encyclopedia* (Ingebretsen, 26). Gradually, the association Frost and Francis shared altered over time from a kind of schoolboyish obsession to symbiotic cordiality to strained divergence.

Early on, Francis, giddy as a paparazzo, records in his journal, "I have not yet met Robert Frost. But I feel his influence in anticipation" (*Travelling in Amherst,* 10). Later, he published the somewhat sycophantic "Robert Frost in Amherst" in the *Springfield Republican* as a "pop-gun fired in [his] private campaign to establish a significant relation with this most significant man in town" (*Frost: A Time,* 49). Other entries preserve his concerns about influence and originality: "Robert Frost has pulled my orbit a little nearer his. He may have pulled me a bit out of my true shape. But the stuff in me is still my own stuff" (*Frost: A Time,* 75). Then, as a man in his fifties, Francis adopts the stance of the resigned realist after a meeting with Frost in Amherst's "Lord Jeff" Inn: "I spent an hour with . . . America's greatest living poet [who] . . . sat in undramatic fashion and chatted with one whom he himself called 'the best neglected poet.' I am the poet whom the editors reject; Frost is the poet who rejects the editors" (*Frost: A Time* 11). Two years after Harold Bloom published *The Anxiety of Influence* (1973), a seventy-five-year old Francis is found still trying to mend a wall between himself and Frost, as if to close the subject for good. "It was natural to compare me to Frost, greatly to my disadvantage," he reveals. But then he insists, "Since my first three books, I have been getting farther and farther away from anything that resembles Frost. So that for years now, I haven't heard anything of that sort" (Phillips-Cubbage, 166). The title that Frost gave Francis—"best neglected poet"—circulated widely, to the degree that it enjoyed prominent display in Francis's *New York Times* obituary in 1987. After decades of living through the "mixed blessing" of their intimate acquaintance, as Robert Phillips called it, even death could not separate the elder poet from the younger (Robert Phillips, 9). Like the wind and the rain, they might remain mingled forever in the similarity of their names, the town they shared, and the landscapes that fed their lives and poetry. For my part, however, beyond emphasizing lesser-known information about Frost and Francis, I wish to sidestep decades of comparison and expend my energies following Francis down a road less traveled.

In spite of decades of prejudice and protean scholarly pigeonholes, I argue that Francis deserves a broader and more environmentally focused examination and that in the age of environmental concern we look at Francis's early and late poetry, fiction, non-fiction, journals, and correspondence and finally read him "in bulk," as Donald Hall indicated we should (Hall, "Two Poets," 123). This proposed "bulk" reading casts Francis's published and unpublished works as a summons to greater understanding and beneficial reciprocity between human and natural worlds. Henry Lyman's *After Frost: An Anthology of Poetry from New England* nudges Francis in this direction, showing that Francis's view of nature's "immensity" translates into "affection," one that treats a "stormy night" like a "familiar, well-loved text" (Lyman, 27–28). Moreover, David Young's introduction in *Longman Anthology of Contemporary American Poetry* depicts Francis as a poet-prospector who hunted for and forged metaphorical relationships among "natural things" such as "toads, cypresses, waxwings, [and] weather" and the "simultaneous matching in nature and language, the two realities of world and word" (David Young, 63–64).

Though there are many ways to approach Francis's writing, I feel that his readers can gain greater access to his work through a consideration of the historical periods of literary development he straddled (modernism and postmodernism) and an application of a hybrid interpretive method that links Francis's life and literature to the earth (biographical criticism, ecocriticism, place studies, and environmental theory). While clear Frostian echoes pervade Francis's work, striking reverberations gesture also toward Dickinson, Whitman, Jeffers, Robinson, Stevens, Pound, and Williams. Francis also offers open-handed jibes at Eliot and Auden. Despite these traces of modernist influence, however, Francis remains the possessor of a fervent wilderness aesthetic, both universal and original. If Pound and Williams serve as icons of imagism, Francis operates as the incendiary of the outdoor imagination. If Stevens strums the modernist maestro's lyre of the abstract, Francis remains the connoisseur collector of the concrete. In approaching Francis's work through an eclectic green lens of cultural history and literary theory, I hope to address two broad but related questions: First, why has literary studies in general not given Francis his due? And second, in the presence of ecocriticism's pervasiveness, why—when scholars have produced similar studies of Dickinson, Whitman, Jeffers, Frost, Snyder, and Berry—has no one invited Francis to take his place in the sun next to them?

Francis's writing contains evidence that he was aware of his marginalized status. In those moments, he strikes back at fate with the double-edged sword of satire. In his bite-sized essay "Anthologists," he employs the vernal icons of a New England habitation to cast genial mockery at publishers who failed to categorize him accurately. In this piece, he likens anthologists to flower gatherers, bees, and earthworms due to their tendency to assemble publications with buzzing capriciousness, their flighty seasonal preferences, and their slow-developing but "profoundly subversive" secrecy. "If critics are aristocrats, writing for the few about the few," he jibes, "anthologists are usually democrats, writing for the many and hoping for a good sale" (*Pot Shots* 163). While it remains a dubious task to assign labels to a writer whose prose delivers such bell-ringing bastinados, the scholarly amnesia and chain-gang inaccuracies trailing Francis into the future of American literature call for remediation. It is time to uproot Francis from the obscurity of the green margin and to plant him where he belongs: in the expanding canon of twentieth-century American ecopoets.[11]

Though for years Francis "resented even being called a poet," over two decades after his death his matchless mode of living and writing requires a specialized but apposite category (*Frost: A Time* 83). In an increasingly diverse and expansive body of ecological literature and ecopoetry, prominent figures have emerged bearing custom-made titles. For example, Greg Garrard lists Gary Snyder as the "poet laureate of deep ecology" and Arne Naess as its "philosophical guru" (Garrard, 21). Lynn White, Jr., called for St. Francis of Assisi to be canonized as "patron saint for ecologists" (White, 14). In addition, J. Scott Bryson mentions how many scholars and readers consider Thoreau "the literary ancestor of practically all ecopoets" and Robison Jeffers the "father of ecopoetry" (Bryson, *West Side*, 120; Bryson, *Ecopoetry*, 7). In "In Memorium: Four Poets," Francis appears to agree, listing Jeffers first and granting him laudatory but nameless immortality: "Searock his tower above the sea, / Searock he built, not ivory. / Searock as well his haunted art / Who gave to plunging hawks his heart" (1–4). In this company, and amid such lofty coronations, Francis seems ill-fitted for such royal distinction, though certainly worthy of much higher regard. But if a title must be given, he should rank as no less than ecological literature's foundling or adopted son. If categories such as "marginalized twentieth-century American ecopoet" seem overly cumbersome, I suggest that we assume an evolutionary stance and begin winding the filaments of such an open canon around a cocoon of twiggy gossamer with Francis, if not at its core, at least very near the action.

Far from an erstwhile sports writer or Frost wannabe, Francis evolved into a poet's poet whose sportive frolic in the fields of language and New England generated a symbiotic cross-fertilization for both artist and region. His hunted-and-gathered corpus has scattered in his hushed wake a wonderfully heterogeneous trove of eroded shards and burnished stones, both collected and uncollected. Like a scattered heap of autumn leaves, his work deserves to be raked in so the next century

can circle and let the colors dazzle curious onlookers in the smoky twi-
light of his passing.

One standout dimension to Francis's ecopoetic explorations is his con-
tinuous revelry in the seasons and elements of the earth. Above all, his
writing seeks to capture the earth's most elusive aspects: the weather, the
untraceable composition of soil, the changing temperature, the unstop-
pable symphony of erosion and precipitation, and the cycle of warmth
and cooling. As an ecopoet, he authored what might be called the sea-
sonal and elemental text. In a letter to Charles Sides, Francis describes
how, for him, poetry results from the slow interplay of organic forces
and factors, as if a poem were a crocus forcing its way through the soil
to reach the sun, or a reservoir of liquid nickel churning in the magma
smith of the earth's belly: "My ideal is that the growing poem itself wants
to fulfill itself in one form or another, and that the poet needs to listen
to the poem in order to help it fulfill itself" (Sides, "Freedom," 2–3). In
The Environmental Imagination, Lawrence Buell highlights the concept
of "the elemental text," extracting his terminology from the governing
structure of Aldo Leopold's *Sand County Almanac*, whose shape takes
its direction from the "traditional four elements—water, fire, air, and
earth" (Buell, *Environmental Imagination*, 227). "The craft of environ-
mental reading," Buell points out, "may jog us in the direction of want-
ing to make ourselves more aware of how phenomena signify and how,
beyond that, even our suburbanized, attenuated lives are subtly regulat-
ed—maybe even constituted—by the elements" (Buell, *Environmental
Imagination*, 251). In a kind of medieval regression toward simplicity,
Francis appears to have spent much of his time writing about embrac-
ing a life centered on the universe's primeval elements as understood by
previous civilizations.

In particular, Francis demonstrated an artistic fascination with the
traditional element of fire—sometimes figuratively, sometimes literally.
If fire constitutes a primal combination of human and non-human ele-
ments, both a conflation and conflagration, Francis's continued inter-
est with it reveals much about how he saw his place in the world and
the world's place in him. His journal records that, previous to his first
book publication, he staged a vespertine sacrificial ceremony of sorts,
the torching of a decade's worth of rejection slips: a total of 641. His
description of the backyard inferno communicates an exhilarating sense
of cleansing catharsis: "[They] blazed beautifully in the wind against the
black, and they kept blazing a surprisingly long time. . . . I thought they
would never disappear. I stood and waited and warmed my hands" (*Trav-
elling in Amherst*, 41). In another entry, he uses the language of fire
to describe the metaphysical transformation he underwent in the act of
writing a poem: "This morning as I lay in bed the conception of a poem

came to me. . . . I lay quietly for the few minutes required for the poem to catch fire. Then after breakfast, I did not drive myself to an eight-hour job, smoldering with impatience and rebellion, but I sat down in my study and burned cozily the whole morning. At dinner time, the verses were done" (*Travelling in Amherst*, 2). Here, the heat in his language suggests that he felt poetry operated like a volcanic eruption or lightning strike, a simultaneous consumption and consummation, something the poet could not stop if he wanted to. Everywhere, Francis's poetry and prose catch the soul dwelling at white heat in the crucible of nature as both flamethrower and furnace.

From 1936 to 1993, the elemental subject of fire flares and flickers in his poems, in luminescent stanzas and titles as brief as sparks: "Fire Warden at Kearsarge," "Bonfire on Snow," "Fire Chaconne," and "Moveless the Mountain Burns." In "Late Fire, Late Snow," aged poet and planet, like flint and steel, collide and ignite an eternal, life-giving torch that fuels both:

> White age supposedly retired
> Above ground or below
> When suddenly a flare
> Of superannuated fire
> Fire—or a late late snow
> After the snowdrops
> After the crocuses
> How the great flakes plaster
> The whole blooming earth
> As if to say I'll show them
> As if to say I'm still above
> For another day (1–12)

In these lines, whose fragmented syntactical ruptures mirror the disconnections between tectonic plates and lightning strikes and snow gusts, Francis writes of the human spirit in terms of the blooming earth's molten magnetic core, and vice versa—always in flux, always interdependent, never finished.

In an earlier poem, "Three Darks Come Down Together," the writer's soul internalizes the pyropoetics of the earth as a spiritual protection against pending destruction. The ends of all human lives, and their many beginnings, huddle tenuously within the weathered pavilion of an unforgiving but purifying natural order:

> Three darks come down together,
> Three darks close in around me:
> Day dark, year dark, dark weather.

They whisper and conspire,
They search me and they sound me
Hugging my private fire.

Day done, year done, storm blowing,
Three darknesses impound me
With dark of white snow snowing.

Three darks gang up to end me,
To browbeat and dumbfound me.
Three future lights defend me.

Into the enduring action of the four elements, Francis infused a fifth—the human. In doing so, he lit not a raging forest fire that soon extinguished itself, but a smoldering coal in the woods whose steady orange glow is only now rising with the intensity of a new sun to warm a weary world.

2
The Influence of Dickinson and Frost

> You see, I've never had to go anywhere to find my paradise. I found
> it all right here—the only world I wanted—here in Amherst, Mas-
> sachusetts.
>
> —William Luce, *The Belle of Amherst*

> His lyric style, casual yet compact, reminded me so much of Rob-
> ert's that until I learned better, I thought my leg was being pulled
> and that Robert Francis was an alter ego Robert Frost had invented
> by slightly altering his last name.
>
> —Louis Untermeyer

WHILE THE GEOGRAPHY AND CLIMATE of New England most certainly influenced Francis in becoming the artist he eventually grew to be, other influences, both topographical and textual, helped shape his authorial ethos. As an American ecopoet, Francis did not simply spring forth out of the ground into the sunlight, untouched by literary tastes and unencumbered by literary history. As an educated man and a devoted reader, he exhibited the tendency to imitate, and rival at times, those writers who wrote before and during his lifetime. In many ways, his poetry reflects the techniques and trends of modernism; at other times, his lifestyle and lyricism hearken back to the nineteenth century, the era that missed claiming him by a matter of years. Amherst's two most well-known poets clearly directed Francis's literary evolution, though both had different roles in fashioning and fertilizing his career. To understand Francis's ecopoetics, then, we must examine his relationship with his most influential predecessors: Emily Dickinson and Robert Frost.

"[T]HAT THIS IS 'AMHERST'":
EMILY DICKINSON AND THE PHENOMENOLOGY OF PLACE

In their 1976 *Pot Shots* interview, Philip Tetreault and Kathy Sewalk-Karcher identify in Francis's writing an admiration for nineteenth-century

41

authors such as Emerson, Thoreau, and Whitman. Though Francis responds that he did admire his nineteenth-century predecessors, he says that he "never tried to pattern [his] life on theirs" (*Pot Shots*, 122). While Francis intended to avoid carbon-copy imitations of his nineteenth-century forbears, he nevertheless drew a kind of interpersonal strength from their historical and cultural legacies.

Of all the nineteenth-century authors whose philosophies and techniques shaped Francis's twentieth-century literary output, perhaps Emily Dickinson surfaces as the writer who exhibited the most pervasive and profound influence on him. For example, consider the following nature lyric:

> With wisdom for our wonder
> The flower and the bee
> In a world designed for plunder
> Have reciprocity.
>
> For the bee takes from the flower
> What the flower wants to give,
> And the bee gives to the flower
> What the flower takes to live.
>
> Though the plan is less than loving,
> Though the ethic is not soaring,
> How far better than our living
> When our loving turns to warring.

Apart from the streamlined grammar and syntax, this poem's subject matter, diction, and meter (especially in the first stanza) might expose it as an edited and reconstituted minor Dickinson poem, or at least a poor imitation. Such an identification would be understandable, given its similarity to a poem like Dickinson's 235: "The Flower must not blame the Bee—/ That seeketh his felicity / Too often at her door." In truth, the somewhat trite pacifist lyric cited above, entitled "Flower and Bee," appeared in Francis's 1944 collection, *The Sound I Listened For.* Macmillan, after rejecting it twice, eventually published that collection when a literary agent, Alan Collins, submitted it on Francis's behalf (*Trouble with Francis*, 49). In September 1944, as fortune would dictate, Francis followed Dickinson's spectral footsteps to Mount Holyoke as an English teacher—a brief period about which he remembered "very little"—but in many more memorable ways he continued to follow her throughout his life, in his poetry and in the way he existed (*Trouble with Francis*, 52). An examination of the Dickinson-Francis connection isolates the phenomenon of one Amherst poet forming the meaning of his life and place by working through the life and place of America's most famous Amherst poet forerunner.

In addition to place associations and literary style, other less obvious details almost mystically connect the lives of "Robert Francis of Amherst, Massachusetts," as David Young dubs him, and Emily Dickinson, whom Philip Booth, in drawing a comparison between the two writers, playfully refers to as "that other Amherst poet" (David Young 11; Booth, 11). For instance, Dickinson's well-known defection from her family's Congregationalist church parallels Francis's amicable break from his father's Baptist faith, despite his father's position as a pastor and a genealogical line in which six of ten Francis brothers entered the ministry (*Trouble with Francis*, 140, 150). Both poet-detractors received ironic religious descriptions from mentors: Thomas Higginson's depiction of Dickinson as "nun-like," and Frost's sketch in a letter to Louis Untermeyer in which Frost paints Francis as a "Puritanical priest" (Phillips-Cubbage, 167). Throughout their lives, Dickinson and Francis preferred a reclusive lifestyle. According to Henry Lyman, Fran Quinn, one of Francis's executors, sprinkled Francis's ashes on Dickinson's grave after Francis's cremation (Lyman). In 2005, Francis ascended from the ashen world of spirits to be reincarnated in a mural alongside Amherst's West Cemetery. The mural, painted by David Fichter, features Francis and Dickinson in close proximity, just above Frost. According to Fichter, several Amherst residents came forward and requested that the mural include Francis's image (Fichter, e-mail to author, June 6, 2007). Both poets suffered from eye trouble in their later years, and neither married; in fact, the long-accepted but disputed scholarly conjecture that Dickinson maintained a single shadowy relationship with an unidentified male suitor echoes Francis's relationship with an unnamed male "Italian in his mid-thirties" in 1958, a claim not substantiated by any other sources but Francis's word alone (*Trouble with Francis*, 211–13). Both writers sought help from mentors (Higginson and Frost), both of whom advised the younger poets to delay publication (*Travelling in Amherst*, 36). And Dickinson's light verse offering, *"sic transit gloria mundi"* (1852) in the *Springfield Republican* matches Francis's breezy but banal "Robert Frost in Amherst" (1932) in the *Springfield Republican and Union* (*Frost: A Time*, 49).

Coincidences, quirks, and trivia notwithstanding, more foundationally artistic and stylistic details forge a strong bond between the landscape, literature, and lives of the belle and "bachelor-recluse" of Amherst, as Francis referred to himself (*Gusto*, 13). To date, though scholars have gestured glancingly to Dickinson's influence on Francis, no one has produced an in-depth exploration.[1] The two Amherst writers' maverick poetics led Donald Hall to label Francis's art as "eccentric as Emily Dickinson's" (Hall, "Two Poets," 120). Robert Bly, in an effort to immortalize the two poets' intertwined lives and literary products, composed a prose poem, "Visiting Emily Dickinson's Grave with Robert Francis," in which the speaker records a version of Dickinson's burial according to the way "Robert says" it happened. Though no more substantial commentary exists, such surface observations point to a largely untapped interest in tying the famous female nineteenth-century poet-recluse to the lesser-known twentieth-century male "semi-recluse" (Sides, "Freedom," 1).

So how exactly did Dickinson influence Francis? And what constitutes the precise nature of their affiliation? If we widen the scope of past assessments and examine the degree to which Dickinson's literary and locational legacy shaped Francis's life and writing, we uncover a variation of poetic influence. This type of influence subverts Bloom's "Anxiety of Influence" and extends Gilbert and Gubar's "Anxiety of Authorship" into the frontier of twenty-first-century American ecopoetics: the notion of eco-influence, or the extent to which one author's geographical and environmental influence overshadows another author.[2] James Cahalan classifies authors such as Francis and Dickinson, those linked to each other and specific locations, as "hometown authors . . . who not only grew up in their hometowns, but wrote about them" (Cahalan, 258). Cahalan sets hometown authors' "home places" apart as "formative," meaning, in the case of Francis and Dickinson, that Amherst fashioned who they were and how they wrote (Calahan, 268). An appeal to Cahalan's concept of the "home place" author, as well as to the cultural geography of Tim Cresswell, guides us toward a deeper understanding of the many ways in which Emily Dickinson's life and location undoubtedly influenced Francis. Read through a phenomenological association with Amherst, Francis's supposed free-standing status as a marginal twentieth-century American ecopoet gives way to the realization that in order to pay homage to his mentor, Francis consciously offered himself as the male inheritor of the Dickinsonian location and legacy. In Dickinson's words, he emerged as one of nature's "Green People" whose "Fainter Leaves" endured to decorate "Further Seasons" ("Nature Sometimes Sears a Sapling," lines 3,5).

In *Place: A Short Introduction*, Tim Cresswell contrasts social constructionist and descriptive approaches to understanding geographical place with what he refers to as a phenomenological approach. Cresswell's phenomenology of place is best understood in conjunction with what he calls "place-memory": the "ability of place to make the past come to life in the present and thus contribute to the production and reproduction of social memory" (Cresswell, 87). Understood strictly, Cresswell's phenomenology of place seeks to "define the essence of human existence as one that is necessarily and importantly 'in-place'" and is "less concerned with 'places' and more interested in 'Place'" (Cresswell, 51). Within the scope of Cresswell's terminology, Amherst proper becomes not so much a physical point on a map as a timeless location whose porous meta-bound-aries ceaselessly expand to circumvent a repository of meaningful human experience continually reproduced in the collective social memory of people and poets associated with it. Seen through the shifting lens of human experience, Amherst, according to Cresswell, constitutes not a bounded place, but ever-changing *Place*.

Aligned with Cresswell, J. Scott Bryson argues that Dickinson engages in the process of "making place" in many of her poems, such as "Some keep the Sabbath." Of the expansive sense of place-consciousness that infuses Dickinson's writing, Bryson adds, "All are making place when they help reorient us within our world, when they re-enter their con-ceptions of the world in such a way that their poems become models for how to approach the landscape surrounding us so that we view it as meaningful place rather than abstract space" (Bryson, *West Side* 5). Together, Cresswell and Bryson suggest that to poets like Dickinson, a place contains no static meaning itself. Rather, all places constitute high-ways, portals, or slowly eroding weigh stations. Place, with a *P*, treats places as mobile nests that are reconstituted each time the occupants depart or arrive.

More than once in her poetry Dickinson addresses Amherst's Place-ness, its shifting locational storehouse of significant human experience. In doing so, she links it to other heavenly and exotic locations. Of this aspect of Dickinson's poetic and environmental compass, Jane Eberwein writes, "[F]or all her awareness of local and global environments, her truest perspective remained more vertical than horizontal, more attuned to speculations on immortality . . . than on Amherst" (Gerhardt, 57). I argue, however, that Dickinson sees the universal *through* the tran-scendent lens of the local, and vice versa. In doing so, she elevates her hometown from the status of a simple geographical place to the level of universal Place. Consider the opening stanza of poem 176: "If I could bribe them by a Rose / I'd bring them every flower that grows / From Amherst to Cashmere!" Here, Dickinson exclaims, "My business were so dear!"; her business being that of using poetry to project her local envi-ronment directly into the realms of global significance. James Cahalan draws our attention to the neologism "glocal" to remind us how home-town authors such as Dickinson show that "every place on the globe is also local" (Cahalan, 251). In Dickinson's poem 241, her sense of poetic "glocality" extends to the cosmic and universal. "What is—'Paradise'," she asks, wondering if farmers hoe in heaven, then follows, "Do they know that this is 'Amherst'—/ And that I—am coming—too—." In one sense, these stanzas constitute Dickinson's efforts to shape social consciousness, perhaps even cosmic consciousness, concerning conceptual definitions of the place where she lives. Lyrically, she rhapsodizes on Amherst in terms of the grander social and geographical phenomenology that Cresswell's framework of Place affords us.

Francis, taking his cue from Dickinson, employs poetry to elevate their shared place of residence to the level of the profound, universal, and mythic. In this mode, he brings Amherst's past to life in a way that reconstitutes, through art, the reading public's collective social place-memories concerning his town's ever-evolving significance. In "On a Theme by Frost," which ends, "Amherst may have / Had witches I never knew," Francis reconstitutes Amherst as an expanding platial storehouse of mythopoetic magic and subversive lore (27–28). In "The Two Lords of Amherst," he zeroes in on the proximity of two Amherst landmarks—the historic Lord Jeffrey Inn and Grace Episcopal Church on Boltwood Avenue. The juxtaposition of these two structures, so externally opposed in essence and function, drew barbed couplet shafts from the satirical rogue's quiver: "The two Lords, Jeffrey and Jehovah, side by side / Proclaim that hospitality lives and Jesus died" (1–2). The speaker goes on to say that a person could "dart back and forth and not get wet in rain" between the "sacred love" offered at church and the "love profane" proffered at the inn"—"[h]ere are the cocktails, here the sacramental wine" (5–7). Between the "holy" and "not-so-holy" hosts, however, the speaker offers no answers, but only questions: "Tell, if you can and will, which is more richly blest: / The guest Jehovah entertains or Jeffrey's guest" (9–12). This deceptively simple twelve-line satire enacts a powerful earth-moving force on existing social and cultural definitions of what these landmarks and place associations mean.

In Cresswell's phenomenological terms, the shifting Place that included these historic landmarks on Amherst's Boltwood Avenue, before Francis wrote his poem, is and will always be different from the Place created after Francis wrote his poem. In fact, in one account, at a public appearance in 1955, "The Two Lords of Amherst" incited such ire that it was torn down from where it was displayed, until an individual identified as "Bill Merrill" insisted that it be restored to its position. As a poem of Place, Francis's "Two Lords of Amherst" mingles the sacred and profane and creates a site of epistemological, sociocultural, and theological struggle for dominance, and in so doing generates an ongoing cycle of reconstitution about what the smaller places within the Place of Amherst mean. At times Francis exerts great effort to reshape the social memory surrounding one of America's most poetic places—and his place in Dickinson's place (meaning her location *and* her stead). A comparative cross-textual reading of Francis's poetry, essays, articles, and non-fiction prose unearths the breadth of Dickinson's geographically determined influence on him.

For instance, a remarkable case of eco-influence surfaces when we compare Dickinson's "Sunset at Night—is natural—" (427) and Francis's "Two Glories." Brad Ricca groups "Sunset at Night" within Dickinson's "astronomical poems" and suggests that, as a set of works constituting a distinct subcategory, such poems are aimed at "testing arenas of doubt" (Ricca, 103). As Ricca points out, in this compact eight-line composition Dickinson pits the laws of the Christian gospel against the laws of nature, negotiating through poetry the interstellar clash between faith and doubt, the natural and the apparently aberrant:

> Sunset at Night—is natural—
> But Sunset on the Dawn
> Reverses Nature—Master—
> So Midnight's—due—at Noon.
>
> Eclipses be—predicted—
> And science bows them in—
> But do one face us suddenly—
> Jehovah's Watch—is wrong.

Compare Dickinson's eight lines to Francis's eight-line "Two Glories," a poem that treats the supposed unnatural phenomenon of a belated Amherst sunset-sunrise, as measured by a heavenly sprung watch:

> Before it set the searchlight sun
> Broke through a narrow slit of heaven,
> And what all day it hadn't done
> It did in one brief quarter hour,
> And all the light it hadn't given
> It gave with all the purer power.
> Two glories fell on us as one:
> The rising sun, the setting sun.

Side by side, these nearly structurally identical lyrics appear charged with the same topographically determined human experience, as if poetically tapping the same platial ore. As paired poetic sentiments, they suggest that a changing attachment to place operates *like* a sunset: similar to similar individuals in similar places (perhaps in different centuries), but never the same way twice. Citing Edward Casey, Lawrence Buell defines this "[p]lace-sense" as a "kind of palimpsest of serial place-experiences," as "not entitative—as a foundation has to be—but eventmental, something in process" (Buell, *Environmental Imagination,* 73). Buell adds, "The concept of place also gestures in at least three directions at once—toward environmental materiality, toward social perception and construction, and toward individual affect or bond" (Buell, *Environmental Imagination,* 63).

   As this pairing of "Sunset at Night" and "Two Glories" demonstrates, an eventmental sense of Place corresponds to an acceptance of a universe

whose chief characteristic remains a high level of instability and rela-
tionality. From Amherst's shifting geographical station in the cosmos,
Dickinson describes the aerial slippage between sun and earth, and
at the same time, the semantic seismography that switches the human
linguistic referents "sunset" and "dawn." In other words, the speaker
moves readers as she moves herself toward and away from describing
the movement of sun and earth. In the space of eight lines, Dickinson's
speaker, *because* she views the cosmos from Amherst's roving geospheric
platform, explodes all anthropocentric definitions of solar locomotion
and illumination: night becomes day, belief doubt, and sunup sundown.
Nature's forces displace a previously privileged human perspective. Reli-
gious doctrine becomes heresy, and scientific law, fallacy. In the twenti-
eth century, Francis's Dickinsonian lyric furthers this cycle of platial and
relational influence, suspending the sunset over Amherst, as it were, by
making Amherst a Place for not one poet, but—with Dickinson and him-
self as mutually gravitating and orbiting foci—for many poets. Interest-
ingly enough, Francis, writing in 1970, remembers his first impressions
of Amherst in 1926, calling it "a good place for sunsets" (*Trouble with
Francis*, 140).

   In addition to more terrestrial and philosophical topics connected
to Amherst, both Dickinson and Francis hover toward the subjects of
immortality, the after-life, and heaven. It is this thematic shift, Jane
Eberwein argues, that drives Dickinson's "later imaginings" into the
realms of the "more abstract" (Eberwein, 239). In "Heaven is so far
of the Mind" (413), Dickinson ruminates on the Place-ness of heaven,
its locational essence in relation to human consciousness, understand-
ing, and physical geography. "[W]ere the Mind dissolved," the speaker
argues, the exact location of heaven could "not again be proved." Dickin-
son's point is that a "vast . . . Capacity" in the human imagination makes
heaven "as fair—as our idea," in reality, "No further" than "Here." In
one sense, Dickinson first dislocates and then relocates heaven, bring-
ing it down to earth, so to speak. In doing so, she reconfigures heaven's
traditional sense of Place-ness and replaces a Biblical notion of spiritual
ascension with one of radical "con-dissension." Heaven is here, she says,
on earth and in our minds. In one of his later, more abstract imaginings
concerning heaven's place, "If Heaven At All," Francis resurrects the
substance of Dickinson's earthly psalm. "Heaven before I die if heaven
at all," Francis's speaker declares, preferring earthly autumn's "heaps of
gold" over "harps" (1, 4–5). Like Dickinson's speaker, Francis's speaker
views the location of heaven as a state of mind, something as spiritually
malleable and moveable as physical clay: "And as for angels, more beau-
tiful, more real / The young, swift-footed, strong, and visible, / Blond or
dark or auburn like the leaves. / Yes, heaven before I die if heaven at all"

(13–16). Seen in conjunction with one another, these poems, divided by time but connected by Place, harmonize two similar poetic arguments for re-placing heaven and earth, both offered up from the changing terrestrial hub of Amherst.

One experimental example, "Two Ghosts," illustrates how Dickinson steered and tutored Francis as a regional writer. "Two Ghosts" consists of a short dramatic dialogue between two immortal speakers identified by the letters "E" (Emily Dickinson) and "R" (Robert Frost). In terms of its impact on readers, this piece embodies Cresswell's notions of the phenomenology of place and place-memory. In doing so, it brings Amherst's poetic past to life within the framework of the author's present version of the same location, thus molding, redacting, and re-structuring collective social consciousness and memory concerning the place that serves as its setting. At the same time, "Two Ghosts" expands the geographically situated conduit through which it passes in order to generate a sense of a newly re-formed Amherst as Place. As a unique genre, "Two Ghosts" resembles other literary works in which contemporary writers bring dead authors back from the grave and fictionalize fantastic, life-after-death accounts of their influential predecessors, similar to Yeats's Swift séance in his play *The Words Upon the Window-Pane.*

Place-centered language pervades the text of "Two Ghosts." A prose description of the setting precedes the dialogue: "Amherst. Dark hemlocks conspiring at the First Church midway between the Mansion on Main Street and the back entrance (the escape door) of the Lord Jeffrey Inn. Between one and two after midnight" (1). For the most part, the dialogue consists of Francis's imagined conversation between the returned spirits of Dickinson and Frost. At times, the two writers discuss specific lines from certain poems, including Emerson's "The Humble-Bee"; Frost's "Directive," "To Earthward," and "Mowing"; and Dickinson's "After great pain, a formal feeling comes" (372) and "Nature rarer uses Yellow" (1086), in addition to letter 368, which includes her "truth like Ancestor's Brocades" observation addressed to Higginson. While the ghostly Frost and Dickinson represented in Francis's dramatic poem can never be taken as the real poets themselves, reading about their otherworldly representations in this way invites audiences to replace former incomplete conceptions of Amherst and its poets with those of another Amherst poet, while re-Placing Amherst proper.

Though perhaps inconsequential to some, Francis's ghost-dialogue nevertheless alters the social consciousness associated with Amherst and its poets in at least two ways. First, Francis involves the ghost of Dickinson and the ghost of Frost in a conversation that borders on aesthetic nitpicking, the subject being the merit of Dickinson's "Nature rarer uses Yellow" in relation to Frost's "Nothing Gold Can Stay." Though he sublimates this connection, Francis implies that Frost's poem (which *does* bear a striking resemblance to its predecessor) was heavily influenced by Dickinson's (64–75). In doing so, Francis attempts to deflect eco-influential connections between his poetry and Dickinson's by spotlighting influences between Dickinson's and Frost's. The more striking—and perhaps audacious—way in which Francis transforms our sense of Amherst as Place is when he has ghost-Emily actually speak new lines of poetry. Literally, as self-appointed keeper of Amherst's literary legacy, Francis boldly invents the poetry he feels Dickinson would have composed beyond the grave. "E" or Emily says, "Sweet the bee—but rose is sweeter—/ Quick his sting—but rose stings deeper—/ Bee will heal—rose petal—never" (20). To this, ghost-Frost responds, "You talk of bees who were yourself white moth" (21). As the clock tower in the vicinity strikes two in the morning, ghost-Emily speaks in poetry again, "But clocks are human—like us all—/ They err—grow ill—and finally fail," to which spirit-Frost replies, "They never lie intentionally" (60–64). Following these inventions and a smattering of gnomic philosophical exchanges between the fictional spirits of Dickinson and Frost about their putative status as "believers," the drama concludes: "E Two angels strove like wrestlers in my mind: / one belief, one disbelief. / R "After great pain"—/ E Oh! / R Emily? Emily!" (92–96). Despite its obscurity (and badly imitated Dickinson poetry), "Two Ghosts" remains a striking example of how deeply Amherst and Dickinson exerted a timeless, haunting influence on Francis. By all appearances, this influence compelled Francis, as the final survivor of the trio of Amherst poets, to compose a work that would continue to define partially Amherst's locational legacy in the memories of writers and readers who would follow.

Francis's prose, in addition to his poetry, exhibits a tendency toward a place-centered Dickinsonian influence. In 1979, Francis authored an article in the *New England Review* entitled "Emily for Everybody" (McLennan, 49). Essentially, this article critiques William Luce's play *The Belle of Amherst*, a dramatic production that Adrienne Rich labeled "specious and reductive" and that Francis called "ironic and bizarre" ("Emily for Everybody," 505; Rich, 177). "Emily for Everybody," while strewn with impassioned but problematic conclusions, nevertheless provides proof of Francis's conviction that he, as Amherst's poetic heir apparent, could safely peer into the "shadows and the silences" through a kind of critical homage and set the record straight about "both the outward and inward Emily Dickinson" ("Emily for Everybody," 510).

In 1980, the University of Michigan Press, in its Poets on Poetry series, published Francis's *Pot Shots at Poetry*, a miscellany of epigrammatic mini-essays that features no fewer than six on Emily Dickinson. Apart from the general category of poetry and poetics, Francis addresses Dickinson more

THE INFLUENCE OF DICKINSON AND FROST

than any other topic in the collection. Many of these terse essayettes measure a single sentence in length and can be cited in their entirety. Their tones range from reflective to witty to outright absurd. In "Nobody—Somebody," which Alan Sullivan calls so "cleverly turned that it is very nearly lyric poetry," Francis makes a prose posy of Dickinson's famous micro-manifesto: "Somebody, hearing that Emily had called herself a Nobody, decided to be a Nobody too—not just any Nobody but a Nobody who really was Somebody, like Emily" (Sullivan, par. 7; *Pot Shots*, 39). In "Computers," Francis reports how someone plugs Dickinson's poetry "through a computer" and how the computer ends up not working "very well" as a result (*Pot Shots*, 174). In "Emily," he describes how Amherst, Dickinson, and by extension himself are self-evidently inseparable and perhaps synonymous in the wake of her death: "In Amherst, when someone leans out a car window and asks the way to Emily's grave, one does not ask, 'Emily who?'" (*Pot Shots* 206). In "E. D." he spins a surreal, hyperbolic biographical sketch for Dickinson in which she receives the Pulitzer Prize, serves as Consultant on Poetry at the Library of Congress, visits the White House, receives funding from Amoco and Exxon, and speaks to *Time* reporters about Whitman's philosophies concerning poetry audiences and her own love of cash (*Pot Shots,* 204). Finally, in "Local Poet" Francis's alter ego, the satirical rogue, declares the phrase "Local Poet" impossibly "ambiguous," adding, "It may mean merely that the poet happens to live in the same locality that you do. Or it may mean that the poet is confined to that locality, unknown beyond it." Using the specific focal locality of Amherst as a case-in-point, the speaker posits that the phrase "local poet" could mean that the "local poet not only lives here but that his poetry lives here too" with "roots as well as blossoms." In the end, the whimsical speaker expands the definition of a local poet to include international figures from all historical time periods: "In this sense Dante is very much a local poet. So is Wordsworth and Emily Dickinson" (*Pot Shots,* 218). In "Local Poet," and other short prose pieces, we do not need to dig too deeply beneath Francis's rhetorical mélange of playfulness and sobriety to discover how he sees the bedrock essence of his locality inextricably intertwined with Dickinson's—thriving and expanding in continuous mutability.

Along with published prose, Francis authored several fascinating unpublished essays about Dickinson that classify him as a self-styled keeper of Dickinson's personal, poetic, and platial legacy. In one essay, "Emily Dickinson: 1965," Francis observes, "To be ED, one should have ED's environment, and that is something we cannot have. And yet I believe that from her environment, from the way she reacted to it, imposing her will on it while accepting it, we can draw some lessons for ourselves" ("Emily Dickinson: 1965," 11). In the engrossing and intensely personal "Emily Dickinson: Her Posthumous Drama," Francis encounters Millicent Todd Bingham on "ED Day," a day in October during which Amherst celebrated its two-hundredth anniversary. During a tour of the Dickinson homestead, Bingham narrates and then points suddenly: "'Lavinia found hundreds of Emily's poems in a box in the bureau that stood *there.*' Mrs. Bingham pointed to the spot where I happened to be standing" ("Emily Dickinson: Her Posthumous Drama," 16). Here, Francis situates himself in powerful and suggestive ways into the town, house, rooms, and very space Dickinson occupied. In this scene, he and the ghost of Dickinson's poetry manuscripts literally share the same place. In "Emily and Robert," Francis gestures toward a phenomenological view of Amherst as a dynamic site, a place whose significance was reconfigured in the universal social consciousness in the passing of two prominent poets. Dickinson "belonged exclusively to Amherst," he argues, while Frost "belonged to many places, in succession and at the same time; but he belonged to no other place more than to Amherst and he belonged to no other place longer" ("Emily and Robert," 1). Noteworthy here is Francis's implicit conception of Amherst as a fluid place-hub from which all other locations derive and toward which all other locations extend. Note, too, that it is Dickinson and Frost who belong to the place of Amherst, and not the other way around. And when they died, Francis suggests, Amherst as Place continued to evolve: "'Strange,' said a friend yesterday, 'not to see Frost around.' Year after year he appeared in the college town of Amherst, . . . with something of the regularity of a migratory bird. . . . After Emily's death, . . . whatever the people of Amherst said, it is safe to say they did not say, 'Strange not to see Emily around.' All the strangeness had already been. She had not waited for death to become a ghost" ("Emily and Robert," 17).

Those who could never be replaced, Francis seems to say, changed the place they called home by arriving, staying, and departing, though Amherst never ceased to change even when Dickinson and Frost ceased to exist.

At times, Francis's self-professed platial connections to Dickinson transcend space, time, and death. Most noteworthy is his account of his stint as a lecturer at the Chautauqua Writers Workshop, when he served as a substitute for an ill David Morton. Driven by angst and ennui, Francis recounts, he once deviated from his tedious lectures on contemporary poetry ("damn stuff") and spoke from the soul:

> Only once was it different, and that was in 1956 when I spoke on Emily Dickinson. . . . I closed my talk by telling of two times when I had felt ED almost physically present. One was in reading Thomas Wentworth Higginson's vivid description of his first encounter with ED in 1870. I read his words. The second time had been on December 10, ED's birthday, when

a few Dickinson memorabilia had been on display at the Jones Library in Amherst. One of her white dresses was in a glass case, looking strangely as if she had just taken it off or were about to put it on. I said this and stopped. Instead of getting up and running out, the women sat unmoving for several moments and looked at me. (*Trouble with Francis*, 110)

In this passage, Francis's words mingle with Higginson's and Dickinson's (note the familiar use of her initials) in a way that restructures our conceptions of Amherst and its poetic significance. This intermingled renewal of lives, language, and locations brings the past to life for Francis's 1950s audience—and for readers today. It transports all who read it from contemporaneous frameworks of social consciousness to dwell, temporarily, in the realms of Place, where Dickinson, through the medium of Robert Francis, returns to register a forceful carol about the impact of her life and locality.

The subject of Amherst's expansive platiality and boundless multiplicity surfaces also in chapter fourteen of *Traveling in Concord*. This chapter, "Traveling in Amherst," begins with meditative descriptions of Amherst's "four cemeteries." This compass-oriented introduction solidifies as well as loosens Amherst's metaphysical boundaries by direct appeals to the readers sense of limits: "But why talk of Amherst in a book based on Concord? . . . Amherst is Concord translated into my own vernacular. The reader is invited to make a re-translation of his own" (*Traveling in Concord*, 145). Throughout the chapter, Francis's geographical descriptions of Amherst flex the supposedly stable boundaries of space and time. As if daubing his metaphysical brush in the paints of a global palette, he describes the invasive architectural influence in the structures of Amherst College as a "wave of the gothic"; the deceased children of "Mr. Medad Dickinson" in the West Cemetery as "foreign as the natives of Borneo or Patagonia"; a "relic chestnut" as "startling and outlandish as a Greek column rising from the sands of Palmyra." Pithily he observes, "Amherst is Switzerland as well as Rome" (*Traveling in Concord*, 147, 153–54). The moment that the local traveler "unblindfolds" himself, Francis writes, constitutes a revelatory action at once distinct from and absorbed within the ongoing Amherstian phenomenon.

Chapter 14 in *Traveling in Concord* offers an oxymoronic conclusion, a dizzying but also grounded description of how Amherst allows one to travel the globe without leaving the city limits. Amherst weather that would "daunt a tourist" constitutes a "traveler's good luck," the speaker finishes, assembling a catalog of global connections: Amherst in "deep snowfall" as a "trackless Arctic"; Amherst during "high wind" as the "Great Plains, yea Siberia itself"; Amherst in "deep fog" as an "island in the Atlantic"; Amherst noons that border on "tropical"; and Amherst Octobers as clear and intensely blue as the "Mediterranean." Amherst's

atmospheric conditions and weather displays have the power, the speaker
asserts, to dissolve the human subject's concrete sense of place, and
replace it with universal Place. "In winter an ice storm will make a trav-
eler rub his eyes to know exactly where he is," Francis writes. "Even in
ordinary moonlight, what building, what tree, what passer on the oth-
erwise unlighted road, is not steeped in strangeness? Quite possibly the
traveler becomes a little strange to himself. What more can travel offer?"
(*Traveling in Concord,* 155). Careful to distinguish between "tourist" and
"traveler" in this final passage, Francis proposes the existence of a platial
transportation on a globally and personally existential scale. Absorbing
his description, readers sense that this worldwide transportation is avail-
able only to those in Amherst who remain precisely where they are.

For some, such a destabilizing concept of Place seems disorienting
and unnerving. After all, aren't all places set in stone or at least on fairly
stable soil? Physically, they are, but not socially, artistically, historically,
culturally, and imaginatively. Even cosmic, oceanic, and arctic places
exhibit a high degree of changeability. "Place as an event," Tim Cress-
well writes, "is marked by openness and change rather than boundedness
and permanence" (Cresswell, 40). "This World," Dickinson wrote, "is not
Conclusion." Before vacating his Cowles Lane apartment and moving
back to Fort Juniper "for good" in August 1975, Francis described the
unboundedness of Amherst in an unpublished essay entitled "Two-home
Francis": "Less than a year ago it came over me that I could and should
be a two-home man. Not Amherst and Florida. Not Amherst and Italy.
But Amherst and Amherst, the two homes only three and a half miles
apart. My logistics are sound. My life is still centered" (Phillips-Cubbage,
153). This notion of one's home as paradoxically centered and center-
less—as a boundless complex of multiple interconnected "centers" where
the human and wild intermingle—runs as a connective thread through
the Francis canon. "Home," he says, on Henry Lyman's WFCR radio
program, *Poems to a Listener,* "is this little house in which I live, and
much beyond it" ("A Poet's Voice").

If twenty-first century readers consider Dickinson's eco-influence on
Francis, they begin to see ways in which Amherst itself struggles to main-
tain its illusory status as merely a geographical place and becomes sifted
into the ever-shifting slipstream of Place as an event. No longer—and
perhaps never—a point on a map, it remains an always-changing com-
position, a daily equation of poetic significance and compounded human
meanings. A swelling of ground, an expanding circumference. To read
Francis's short essay, "Emily and I," within the context of Cresswell's
place-memory and phenomenology of place is to divine how Francis saw
himself caught up in the cosmic ring of that growing circle of earth that
was, is, and will be Amherst:

The year that Emily Dickinson was born (1830), my father's father, Daniel, was a lusty young man of twenty who in that very year left Ireland for America by sailing ship. When Emily was fifteen, Daniel entered the Harvard Medical School from which he graduated two years later. When Daniel died in 1867, after twenty strenuous years as a country doctor in Nova Scotia, Emily was at the peak of her poetic power or a little beyond, and my father was one year old. When Emily died in 1886, my father was a lusty young man of twenty, and I was born fifteen years later. (*Pot Shots*, 60)

There is some force greater than death, Francis suggests, that will always link him to the poet who could not stop for it.

In a handful of surviving obituaries, reporters link Francis to Frost, rather than Dickinson, as one of America's "best neglected" poets. Inadvertently, these obituaries distance Francis from his eternal weld to Dickinson's (and his) most significant town by misprinting the title of his published journals as *Travelling in America* rather than *Travelling in Amherst*. In 1987, the year Francis died, *Newsweek* slotted the announcement of his passing in a snug corner of a single page, where it lay—displaced and out-of-place—in the proximate shadow of blurbs about Tammy Faye Bakker's new makeover, a David Bowie–Tina Turner Pepsi commercial, Courtney Cox's upcoming roles on *Family Ties* and the *Masters of the Universe* movie, and Bo Jackson's bid to play for both the L. A. Raiders and the Kansas City Royals. Nothing is said of Francis's rightful place in American and Amherstian literary history. In the *New York Times* obituary, however, despite another misprint, readers catch an evanescent but enduring glimpse of a local poet who in a flash of poetic excellence and a glimmer of eco-influence faded from the earth at Cooley *Dickinson* Hospital.

AFTER APPLE PEELING

While it is easy to speculate on how Robert Frost influenced Francis, it is perhaps more interesting at this historical juncture to consider how Francis influenced Frost and how he may re-shape Frost studies. Textual and stylistic echoes aside, one moment in their lives—the infamous "Apple Peeler" incident—sheds a truly revealing light on the nature of their curious relationship and how Francis's sexual orientation created a moment of conflict. Apparently, the "Apple Peeler" microdrama affected Francis so deeply that he immediately captured a blow-by-blow account of the exchange in writing and only published it later as *Frost: A Time to Talk*[3] at the suggestion of Lawrance Thompson, co-author of *Robert Frost: The Later Years*. As if arranged around the moment of this molehill of misunderstanding, *Frost: A Time to Talk* features as its frontispiece a rare photograph Francis took of Frost (with Frost's permission) on the day the elder poet lost his cool: October 30, 1956. Though the event reveals much about Frost's ego and paranoia, Francis, with characteristic understatement, refers to the unexpected kerfuffle as merely "something remarkable" (*Frost: A Time*, 37).

In 1953, Francis's poem "The Apple Peeler" appeared in Rolfe Humphries's *New Poems by American Poets*. In 1956, it was reprinted in Auden's *Faber Book of Modern American Verse*, which became *The Criterion Book of Modern American Verse* in the United States—an editorial kindness Francis coolly repaid with the satirical "Tribute to W. H. Auden and His Vocabulary" in *The New Yorker* the same year! In "The Apple Peeler," Francis's speaker pans back from the action of an old man peeling an apple to locate all human activity in the more global context of the earth's vast ecosystemic processes: "Why the unbroken spiral, Virtuoso, / Like a trick sonnet in one long, versatile sentence? / Is it a pastime merely, this perfection, / For an old man, sharp knife, long night, long winter? / Or do your careful fingers move at the stir / Of unadmitted immemorial magic? / Solitaire. The ticking clock. The apple / Turning, turning as the round earth turns" (1–8). With the two poets bandying Auden's book back and forth casually at Fort Juniper, Frost, in a strange moment of mentorly misprision, made his accusation through a series of strong hints: that Francis's "The Apple Peeler" was a coded slam on Frost's "The Silken Tent." To Francis's astonishment, Frost insinuated that in publishing "The Apple Peeler" three years prior, Francis had used poetry to disparage Frost behind his back for a putative lack of authentic emotional inspiration and a dependence upon linguistic gimmicks for churning out single-sentence "trick" sonnets (*Frost: A Time* 38). According to Francis's account, though Frost was not at first entirely swayed, Francis convinced the older poet of his error, after which they dropped the subject and resumed their cordial association.

What Francis did not know is that, three years previous to the confrontation, Frost had enacted secret revenge on him—also with a poem. In an apparent rage, Frost had scrawled the scurrilous and homophobic "On the Question of an Old Man's Feeling" to punish Francis, if not in public, at least in his own mind, for the perceived slight. With coarse diction and language both ribald and pornographic, Frost's chicken-scratch "pome" uses Francis's love for praying mantises as a vehicle for bashing his homosexuality. In aggregate, the most remarkable aspects about the "Apple Peeler" episode are these: that Frost used poetry as an outlet for revenge; that he waited three years after writing the poem to confront Francis; and that Francis did not learn of Frost's poem or spite until after

Frost's death. Even after Francis discovered the extent of Frost's actions, he responded in a way that invites further curiosity and admiration.

In August 1976, following the publication of the Thompson-Winnick Frost biography, *Robert Frost: The Later Years, 1938–1963,* Professor Jac L. Tharpe, at the University of Southern Mississippi, solicited an essay from Francis for an upcoming book on the "real" Robert Frost. Francis agreed, his stated motivations being to illustrate accurately both Frost's "vindictiveness and his friendliness and generosity" (Letter to Tharpe, September 14, 1976). During the year-long negotiation process, Francis attempted to publish his essay in periodicals before having it included in Tharpe's collection. Having failed to debut "Frost as Apple Peeler" in *New Boston Review* and elsewhere, Francis consented to have it included in Tharpe's book—only to withdraw it without explanation at the last moment. We can only speculate as to why Francis retracted his essay at so late a date after preparing so long for its publication. In his letter to Tharpe, he provides no reason: "After much thought I have come to the conclusion that it will be better for this essay not to be included in your forthcoming collection of Frost essays" (Letter to Tharpe). Nor does he explain the apparent about-face of sentiment that led him to print the article in the inaugural issue of *New England Review* in 1978.

In "Frost as Apple Peeler," Francis clarifies his involvement in what he calls Frost's "darker phases." From his unique perspective, Francis reveals the "jealousy and vindictiveness" that drove Frost to punish him, but he also claims that in keeping the hurtful composition from his sight, Frost graciously intended him not to "suffer" ("Frost as Apple Peeler," 32). Francis also discloses that while persistent scandalmongering biographers like Thompson (whom Francis classes as "vigorous and dynamic") wanted to "beard the lion," he had never had "any impulse to do so" and had published *Frost: A Time to Talk* as an afterthought only, having recorded his conversations with Frost for private use without any intention of making them public ("Frost as Apple Peeler" 35). Francis's conclusion articulates how he put the episode to rest. "If, in punishing me, Frost wrote lines too vile to see print, was he not actually punishing himself?" he asks. Confident in his feelings, Francis records that his admiration for Frost was not compromised and that he found himself "rather amused" by the conflict. "The final irony," he concludes, "is that the poem Frost thought I had written against him is against nobody. It disparages no one, neither apple peeler nor poet" ("Frost as Apple Peeler," 39). Thus aesthetics met ethics. A marginal American ecopoet discarded the rinds of an artistic friendship momentarily gone sour and left the world with the ripe core of justice.

Before, during, and after all the apple picking and apple peeling, Francis's record of Frost's mentorly role reveals much about how one of

America's most lauded poets helped fashion an upstart writer. Time after time, Francis religiously incorporated Frost's revisionary suggestions in his poetry. In response to an early draft of Francis's "Cloud in Woodcut," Frost called the word "high" too "throaty" in the line "Teach your knife high compromise" (*Frost: A Time* 69). In Francis's *Collected Poems*, the line reads "teach your knife to compromise" (*Collected Poems*, 6). In Francis's autobiography, a letter from Frost occupies an entire page: "I am swept off my feet by the goodness of your poems. Ten or a dozen of them are my idea of perfection. A new poet swims into my ken. I can refrain from strong praise no longer" (*Trouble with Francis,* 19). In his autobiography, memoir, and Phillips-Cubbage interview, Francis proudly quotes the letter Frost wrote to Louis Untermeyer in which Frost calls Francis's "opinions . . . no pushovers" and wishes that Francis "might prevail" (Phillips-Cubbage 158). Early on, we see in Francis the agreeable acolyte basking in the glow of Amherst's Apollo. Later, he alternates between maintaining a cool distance from and generating a nostalgic gravity for his wise captain's company.

That Francis never succeeded in putting Frost behind him is manifest in the abundance of published and unpublished writing he produced about Frost after Frost died. Up to his death Francis continued to muse on the subject. In his autobiography he writes of having known "three Frosts": the man "in and behind the poems," on "the stage," and "sitting face-to-face." Later, he expands this list to include a "fourth" and "fifth" Frost—the aggressive career-forging one in the letters to Louis Untermeyer and the one Francis "wanted Frost in his later years to be," someone "able to forget his enemies, real or imagined, to take less seriously the expectations of his audience for entertainment, and to accept fully his great achievement and acclaim without the insatiable desire for more and more" (*Frost: A Time,* 99). In the final "Frost Today" chapter of *Frost: A Time to Talk*, with Frost safely behind the curtain of death, Francis challenges many of Frost's "pet notions" and metapoetic metaphors: the "playing tennis with the net down" criticism of free verse; the "ice on a hot stove that rides its own melting" observation about spontaneity, and the potato as metaphor for two kinds of realists: those who want the potato "with all its dirt clinging to it" and those who want it "scrubbed clean." In response, Francis conceives of "other games than tennis that can be played on a tennis court, games in which a net would be irrelevant and even a hindrance, yet games fully as exacting as tennis." Unflagging in his replies to Frost's ghost, he argues that "spontaneity sometimes has to be labored for" and that "this is not a world in which we choose between dirty and clean potatoes; this is a world in which we need and welcome both" (*Frost: A Time,* 84–86). In a single sweep of memory, Francis lists his top twenty Frost lyrics as "cut and shining gems" and as "momentary stays against confusion," then turns around and eviscerates Frost's famous expression: "Why 'momentary,' Robert? Why aren't they permanent stays against confusion, enduring as long as the poems endure?" (*Frost: A Time*, 98). Though Francis, a lover of polysyllabics, makes his admiration for Frost no secret, as a technical parting shot, he criticizes Frost's "fondness for monosyllables" in poems like "A Servant to Servants" where a monotonous "crumbling" and "plodding" effect became "too much" for Frost at times (*Frost: A Time*, 96–97). At last, in these desultory remembrances the best neglected poet no longer neglects his best opinions.

In certain places Francis hints that Frost assumed a more mentorly role than Francis desired, and that Frost did not fully grasp Francis's economic hardship. With a trace of annoyance, Francis tells Jay Parini that Frost habitually appeared "unannounced" for visits at Fort Juniper (Parini, 387). Elsewhere, Francis captures Frost at "less than his best"— the day in 1959 that Frost appeared with Alfred Edwards of Holt Publishing to offer to publish Francis's latest manuscript—only to have Francis, after fifteen years of remaining unpublished, anticipate the offer and turn them down because Wesleyan University Press had already accepted it! *"Oh, why did you do that?"* Francis remembers Frost exclaiming and, *"Too late, too late,"* as the senior poet left Francis's home for the last time (*Frost: A Time*, 47). Any aspiring unknown writer who has had the luck (or misfortune!) of publishing his or her work with the help of a more successful writer can only look with a blend of astonishment and admiration at this one act of candid refusal. By itself, this incident casts Francis as a truly independent spirit.

Frost exasperated and entranced Francis, even after death. Years before the Holt incident, Frost dumbfounded Francis with the following casual question: "What do you do when you're not actually writing?" Too shocked to respond, Francis realized Frost had no idea how "vastly different" their lives were: Frost's all "laissez-faire except for the poetry" and Francis's filled with "[m]arketing, cooking, dish-washing. Washing, ironing, mending, bed-making, floor-mopping. Gardening, grass-cutting, leaf-raking, snow-shoveling. Storm windows off and screens on, screens off and storm windows on" (*Trouble with Francis*, 89–90). Francis's laudatory unpublished essay, "Emily and Robert," praises Frost as an artist "at home wherever he was" ("Emily," 3). But Francis also criticized the unwieldy grammar of Frost's prose, and in his published essay "Frost as Mugwump," he twits the contradictory elder poet for living with his "mug pointing one way and his wump the other" ("Emily," 3; *Pot Shots*, 30). Francis's "On a Theme by Frost" and "In Memoriam: Four Poets" celebrate the witches of Coös and Grafton and the granite, snowy Frost as a worshipper of the "God of Flow," but in the short essay, "The Robert

Poets," Frost remains conspicuously absent among Herrick, Bly, Burns, Creeley, Browning, and others, as if Francis wished to forget his presence completely (*Pot Shots*, 187). In some ways, a single anecdote characterizes the Frost/Francis legacy. In 1975, Robert Phillips accompanied Francis to a reading at New York's 92nd Street Poetry Center. As Francis approached the pulpit to read, he spied a portrait of Frost hanging in the hall. "Even here," he sighed dejectedly into the microphone (Robert Phillips, 9). This *doppelgänger* exasperation extended into Francis's later years, though not everyone shared the conclusions, or inconclusiveness, that he maintained.

Like snapshots, rare documents and scraps left behind weave a multifaceted narrative of how various individuals perceived the two poets. Later in life, when questioned about Frost's influence on his writing, Francis told Philip Tetreault and Kathy Sewalk-Karcher that Frost was "too special, too exceptional to stand as a representative writer." He added, "He and I were too far apart in achievement and status for me to have had any feeling of envy. It was enough to feel the tonic of his power" (*Pot Shots*, 129). Still, we can only wonder how Francis must have ground his molars after he visited Ms. Dmytryk's elementary-school class and received the following mimeographed smiley-face HAPPY GRAM: "Dear Mr. Francis," Ms. Dmytryk writes, "[T]he children . . . thoroughly enjoyed your visit a great deal. Paul brought in the poem, 'The Axe-helve.'" But another student, perhaps at the high-school or university level, derives from a meeting with Francis an incisive commentary on which of the two Roberts embodied the persona of the great American poet. This student depicts Frost as a once vital, but middle-of-the-road writer "in a small apartment near the Charles and close to Harvard" who "sits in his undershirt and speaks . . . his poems" for "10,000 a year or whereabouts." Frost's lifestyle and literary output, to this student, grants him not legendary status but a "pass for vagrancy and a blank check for convention," a former bright star who has turned to "character acting and exhibitionism." To at least one person, then, it was Francis who succeeded in getting something gold to stay.

Like fire and ice they were. Fifty years later, still circling something like a star in Frost's cosmos, Francis, as haunted as ever, composed "For the Ghost of Robert Frost" and "Two Ghosts"—the aforementioned dramatic dialogue/séance in which the ghosts of Frost and Dickinson return to haunt Amherst and swap new lines and ideas about poetry. The "[h]alf ghost, half natural phenomenon" to whom the speaker in Francis's "Old Poet" refers remains significantly ambiguous, at the same time both biographical and autobiographical: "A little cool he seems and sinister. / White, utterly white, his bardic beard and hair" (3, 8–9). Whether the subject of "Old Poet" is Frost or Francis or both, the speaker drapes the poet in descriptors geological, atmospheric, cosmological, and global, enfolding the Hegelian Francis-Frost struggle for identity within the natural processes of the earth. "[I]n spring he comes drifting down / Into the summer-tourist traffic lanes," the speaker opens (1–2). Then, the elderly figure shambles "south," and the women who watch him through "telescope" and "field glass" travel "east" (5–6). "Old Poet" concludes with a central question "A wanderer so far from his arctic mist, / Mortal and fated and melting as he must—/ The wonder is, the wonder is, how long he will persist" (7–9). Two poets. Both Roberts. Same initials. One town. Who ended up haunting whom remains a question for future centuries.

"Critical readers of my poetry," Francis notes in a 1949 journal entry, "notice first the influences" (*Travelling in Amherst*, 66). So which writer exerted the greater influence on Francis? Dickinson or Frost? Francis's journals and other remaining bits of correspondence indicate that both predecessors steered his literary development. In fact, on occasion in private conversation, Frost and Francis often reflected on the merit of Dickinson's work and, in particular, the power she exerted over them as poets, if any.[4] Standing outside Fort Juniper on a spring afternoon in 2007, I asked Henry Lyman about the Frost/Francis connection. Emphatically, Henry spoke to the space between his hands. "I say again," he said, "when I read Francis, I read New England. Not so with Frost" (Interview with author, May 18, 2007). While disagreements over the question of Francis's influences will most likely go unresolved, if we consult Francis's opinion in the trail of writing he left behind, we can see that, in his mind at least, it was not so much the bard but the belle of Amherst who co-authored his letter to the world.

3
Sex, Gender, and the Rural Erotic

> The reason for the existence of three sexes and for their being of
> such a nature is that originally the male sprang from the sun and
> the female from the earth, while the sex which was both male and
> female came from the moon, which partakes of the nature of both
> sun and earth.

—Plato, *The Symposium*

AS AN AMERICAN ECOPOET, FRANCIS constantly undercut overly simplified egocentric categories that tend to define gender, literary genre, and species. His world was a world of hybrids, Indian summers, and transitional terrain. As a gay artist living and writing through the first eight decades of the twentieth century, he became increasingly sensitive to the ways in which sociocultural prejudice and polemics, environmental oversimplification, and literary parameters acted as destructive confinements. Quietly, he used language and literary creation to chip away at "either-or" models of thinking that misrepresented sexuality, gender, genre, and the environment. For example, his life as a gay poet unwilling to choose between strict, culturally defined categories of "male" and "female" sexuality is matched by his prolific mastery of the prose-poem, a form of mixed textuality. His dissident's assault on harmful sexual and literary hierarchies parallels his desire to equalize "human" and "non-human" organisms, particularly in the way he lived and wrote at Fort Juniper on a part-natural, part-man-made boundary that joined civilization and wilderness. Two books—*Gusto, Thy Name Was Mrs. Hopkins* and *A Certain Distance*—show that Francis's brand of American ecopoetics sought for unification, rather than division, and that he wrote in order to strengthen the natural affinities between sexuality, literary genre, and environmental preservation.

THE GENDER OF GENRE

If genre functions as an analogue to gender, then Francis's reflective *Gusto, Thy Name Was Mrs. Hopkins* qualifies as a gender- and genre-bender. *Gusto's* subtitle, "A Prose Rhapsody," forecasts the flight-of-fancy warping of genres it promises to deliver to readers. In *Gusto's* opening sequence, Francis acknowledges the hopscotch trajectory his piece will follow: "If I'm going to write about Mrs. Hopkins, I'll have to ramble, a little like her rambling, very rambling, old white house. It will be a rhapsody, and rhapsodies are not divided into neat sections" (*Gusto*, 8). He also situates his rhapsody in the context of an environmentally sensitive and cosmos-oriented paradigm: "Mrs. Hopkins's house demonstrated perfectly that celestial phenomenon known to astronomers as the Precession of the Equinoxes. Gradually during the past two thousand years, the heavens have slipped a cog, so that when the sun is officially in Scorpio, it is now actually in Sagittarius" (*Gusto*, 11). Here and throughout his piece, with respect to all things textual, cosmic, natural, and interpersonal, Francis implies that to acknowledge that the universe is not completely where it seems to be is to know that things are the way they should be. As the author of this slim rhapsody, he appears primarily interested in the metaphysically nebulous, the cosmically out-of-whack, and the biologically indistinct. Though in many respects *Gusto* appears structurally sound, as a literary text it features significant rifts and nuances that shade even its author's awareness of the social construction of gender in the context of an environmentally sensitive world view. As a text its gender focus is centered in its eccentric and conservation-minded central figure, Mrs. Hopkins, who, Francis points out, once led a public crusade "before the Board of Selectmen to plead for the life of a condemned tree" (*Gusto*, 7).

Though they don't state it directly, *Gusto's* irregular rhythms recycle and mull over the socioenvironmental construction of gender. In a passage that Francis composed at a time when *Gusto* remained unpublished, he reminisces about his reasons for leaving Mrs. Hopkins's home. His words and totemic time frame recall the terminus of *Walden*:

> I am not sure I remember just why I left Mrs. Hopkins after two years and two months. . . . Our relation had been a true symbiosis. But it may be that true symbiosis with Mrs. Hopkins was something I didn't really want, at least any longer. She had bidden me be a man and had helped me to become a little more of one than I had been. I may have felt that being under her roof and supervision (however restrained) took away more from my manhood than she was able to add. So I left. But came back often to call, was invited into her pewter corner and called Robert. (*Trouble with Francis*, 207)

Here, Francis's philosophical view reverberates with subliminal contradictions and implications. Like Thoreau in *Walden's* conclusion (which also functions as a beginning of sorts), Francis describes the arrived-at moment as inimical to further personal progression, though something he yearns to experience eternally. In much the same way that Thoreau achieved greater unity through separation, Francis fosters a greater sense of "manhood" in the presence of a domineering wife, though not his own. Just as Thoreau became everything he was by immersing himself in everything he wasn't, Francis finds his sense of self bolstered by his most opposite other. As Elinor Phillips-Cubbage observes, *Gusto* "reiterates [Francis's] inner drive to mature into a man" (Phillips-Cubbage, 39). Phillips-Cubbage's summary, however, excludes the degree to which *Gusto* troubles traditional concepts of manhood and womanhood, the way that, as a literary text, it revolves on the socioenvironmental axis of man/woman as eventmental work-in-progress. After finishing *Gusto*, the most readers can say is that Francis becomes more of a "man" because of a "woman."

Whether Francis explicitly intended it or not, *Gusto, Thy Name Was Mrs. Hopkins* depicts gender as an event rather than a state. The fugue-like nature of the man/woman phenomenon, in other words, is always "happening." As we read *Gusto*, polarities such as male/female, man/woman, and masculine/feminine begin to disintegrate, switch places, and ultimately fail to function as dualistic referents. For instance, he uses the phrase "man to man" to describe the way Mrs. Hopkins talks to the local chief of police during an interview (*Gusto*, 25). In his autobiography, published seventeen years before *Gusto*, Francis uses the same phrase—"man to man"—to depict the way Mrs. Hopkins's "managerial and maternal instinct" would come into play when she would thunderously sound the merits of an aspiring actor or young poet: "no holds barred, having once got him into her 'pewter corner' over a cup of tea (perhaps spiked with rum); lecture him, work on him till he was in tears" (*Trouble with Francis*, 204). Though Francis's tone at times remains elusive, his language reorients readers to a socioenvironmental paradigm in which the male figure assumes the more passive, traditionally feminine role as object, and Mrs. Hopkins, the dominant female figure, assumes the more traditionally masculine role. Such textual triggers involve readers in the process of beholding gender as an unfinished product in stages rather than *in situ*, an action endlessly recurring, like clay on a potter's wheel—spun, shaped, and broken down to be reformed again. Though she doesn't express the sentiment directly, Gertrude Stein (whose reading Francis and Mrs. Hopkins attend, as if on a date, leaving behind Professor Hopkins) centers her address on the notion that "writing must go on," which Francis pans as a "reasonable statement" (*Gusto*, 28). The texts of gender and nature, he demonstrates, must also "go on" or else cease to be what they are—and what they are always becoming.

At times, *Gusto* depicts the socioenvironmental construction of gender as the sudden result of gentle and violent clashes. In one pivotal episode,

Francis finds the molecular structure of his "manhood" rattled when Mrs. Hopkins, his superior "feminine" other, employs him as a "polite 'passer of cake'" at her annual tea party, which, Francis writes, "went ill with the man in him." He grumbles, "Mrs. Hopkins, like the unregenerate woman that she is, bids me be a man—and does everything in her power to keep me from it" (*Gusto*, 17). In this instance, "manhood" is something Mrs. Hopkins ascribes to Francis, as a result of her *acting upon him*, as if gender were something she could pass him—and which he refuses at times—like cake on a plate.

At his rhapsody's apogee, Francis describes a "jolt" that alters the once "placid relationship" between himself and Mrs. Hopkins. He writes of this "inevitable" moment that "she had bid [him] be a man all right and had not minced her words," in reference to the time that Mrs. Hopkins requested that he start smoking cigarettes as a means of acquiring the "fear-inspiring element" that, according to her, Francis lacked (*Gusto*, 17). During a ride in Mrs. Hopkins's Model A, Francis told her that, contrary to her wishes, he had stopped smoking. On arriving at her home, she suddenly "stepped up" to him in the kitchen and slapped him! His response bears his hallmark placid philosophical tone. "At my birth (according to my father)," he finishes, "the attending doctor slapped me to start me breathing. Mrs. Hopkins's slap . . . also started me breathing. Outwardly I accepted her rebuke like a gentleman. Inwardly I was exultant, exhilarated. . . . In spite of herself she might make a man of me yet" (*Gusto*, 17–18; *Travelling in Amherst*, 20). At the risk of attributing unintended significance to Francis's account, it appears as if he narrates a direct transfer of transformative energies between himself and Mrs. Hopkins, as if he, the gender newborn-in-embryo, were dependent on his host mother to determine his masculinity/femininity. Mrs. Hopkins slaps gender into Francis as if swatting a mosquito.

In *Gusto*'s dynamic drama, however, the superiority that one gender exerts over another frequently changes hands. According to the language in his account, paradoxically, Mrs. Hopkins's power becomes discernibly diminished while at the same time wielding an even greater potency to alter Francis's personal autonomy. In his journals his recollection varies only slightly in terms of details. In one entry (February 2, 1933) Francis relives Mrs. Hopkins's bringing "all her guns into play" against him, while he insists he was "polite, kindly, but obdurate." He adds, "[M]ore important, to both of us was the question whether or not I was to do something on her say-so. Whereas she struck blindly, . . . I was perfectly controlled and thus will have no apologies to make. When I left the house, I felt buoyant. It was as if an evil enchantment had been broken. I felt like singing. Nothing that has happened recently has bucked me up more than this. Having declared my independence I feel more of a man" (*Travelling in Amherst,* 20). Later, Francis records the beneficial aspects of his land-lady's antithetical and slightly hostile force: "Mrs. Hopkins has helped me, but not altogether in the ways that she probably thinks she has. She has helped me by rubbing my fur the wrong way and thus sharpening my awareness of myself" (*Travelling in Amherst,* 24). In that they craft Francis and Mrs. Hopkins as neither completely masculine nor feminine in the traditional sense, these and other passages in *Gusto* blur the sup-posed lines of demarcation that divide "man" and "woman." Instead, they posit the socioenvironmentally engineered notion of man/woman as a rolling collage of dominance and submissiveness.

Historically, *Gusto* gestures toward Thoreau and his gender identity and attitudes. In *Gusto*'s afterword, Gordon Lawson McLennan includes Walter Harding's classification of Francis as the "Thoreau for our time" (McLennan, 45). In many ways, Harding's observation rings true, partic-ularly in passages wherein Francis not only associates his situation with Thoreau's in *Walden* but also fuses aspects of gender and the natural world in an effort to re-write the Thoreauvian legacy in a way few nature writers have. Francis emphasizes that he lived with Mrs. Hopkins for "two years and two months," the "exact length of time Henry Thoreau lived in Walden Woods," though Francis stresses that between himself and Thoreau the "differences were greater than the similarities." Francis declares outright that Thoreau "went to Walden to get away from women, including his mother, a redoubtable woman with perhaps a touch of Mrs. Hopkins about her." Further, Francis claims the "Thoreau household was pro-woman" and "Walden was strictly anti-woman," insisting that "Henry had the best of both worlds, though he would have died before admitting it." In order to distance himself from Thoreau, Francis takes a firm stance as someone who is not anti-woman: "As for me, I was scarcely escaping from Woman when I moved in at Mrs. Hopkins's. Whatever it was I was escaping from, it was clearly not that" (*Gusto,* 19). Whether Francis classes *Walden* accurately as anti-woman, either figuratively or literally, presents a point for vigorous debate. Still, based on Francis's writing, who and what he was in terms of gender stemmed directly from the presence of "Woman." For instance, as if transcribing the contents of his subconscious in his microessay "Goddess," he ruminates on Robert Graves's depiction of the mythical "White Goddess" as a "bloodsucking vampire," then concludes, "Some scholars have questioned the existence of the White Goddess. I fear she is all too real" (*Pot Shots* 55). In the texts he authored, Woman at least partially made Francis, fashioned his nature, almost as certainly as Christian doctrine claims that God made Adam and Eve.

One final textual curiosity in *Gusto* provides a means toward approach-ing an additional strain of Francis's nature prose/poetry: the rural erotic.

While its inclusion in *Gusto* might strike readers as quixotic and tangential, as a thread in the text's tapestry it nevertheless communicates a sense of Francis's interest in literary expression as a mode of exploring the presence of the erotic in the natural world. Midstream in the rhapsody, Francis drops in what he remembers as one of Mrs. Hopkins's "favorite" limericks:

There was a young lady from Renus
Who went to the ball dressed as Venus.
The host said 'twas rude
To come in the nude,
So they brought her a leaf from the greenhus [*sic*].
(*Gusto* 31–32)

Several aspects of this text-within-a-text invite us to read them through the filter of the rural erotic. In assessing Mrs. Hopkins's humor as a preface to the limerick, Francis writes, "If Mrs. Hopkins was earthy, there was usually a touch of elegance to her earthiness" (*Gusto* 31). In employing such a polysemous geocentric descriptor to label this limerick's focus, Francis forges a metaphysical link among the environmental, the erotic, and the masculine/feminine.

In the limerick itself, a part of earth's nature—the leaf—masks the sexuality of the human Venus figure, though incompletely. In one sense, the leaf functions as a penile cover and prosthesis, appropriated from earth's flora, both a virile extension and a badge signifying a lack of inseminating power. It is imposed on the human sexuality of the female figure who, once partially masked and transformed, becomes marked with both masculine and feminine erotic components since a single leaf may more completely conceal a man's sexual dimensions than it can a woman's. Here, the rural erotic assumes an ambiguity. Nature only partly obscures the erotic object that nature itself has displayed. Nature, then, because it acts as "coyote-trickster," as Daniel Spencer argues, performs the dual function of concealing *and* revealing, prohibiting *and* inviting, defining *and* confusing. As Spencer points out, "The world resists being reduced to resource and emerges . . . always problematic, yet always potent with possibilities for negotiating the tie between meaning and bodies. . . . Metaphors of nature such as Gaia or the Native American use of coyote emphasize that nature cannot be controlled, and is always full of surprises and new possibilities" (Spencer, 99, 102). Textually, Mrs. Hopkins's limerick mutes the buried rhyme (penis) that readers expect—but do not receive—in the concluding line and produces a hybrid figure of androgynous rural eroticism: male/female, Adam/Eve, Adonis/Venus. The figure of the "greenhus" (greenhouse), the anticipated erotic rhyme's substitute, also represents the degree to which humanity attempts to contain, and grow, the natural and erotic in a state of artificiality.

In the same way that Mrs. Hopkins's text establishes erotic expectations and then defers and transforms them, Francis's writing of the rural erotic provides a revisionary approach to literary explorations of *eros* in the *bios*. In doing so, he slowly surprises those who read his last work of nature prose/poetry, *A Certain Distance*, with the figure of the male Venus.

THE ECOLOGY OF EROS

In *Painted Bride Quarterly*'s special Robert Francis issue, Fran Quinn, the guest editor, explains how, given Francis's untimely death on July 14, 1987, the staff was forced to transform their "tribute" into a "valediction" (Quinn, 5). This *PBQ* issue contains a visual and linguistic potpourri generated by Francis and other writers: poems-in-process, rare photographs, criticism, biography, nonfiction, and woodcuts by artist Wang Hui-Ming. On the page opposite David Graham's article, "Millimeters Not Miles," readers encounter a black-and-white print of a Wang woodcut. Elsewhere in the issue, Wang explains that this woodcut served as a response to Francis's request that Wang write an autobiography. Put off by the "half-truth" nature of autobiography, Wang opted instead to author an autobiography in "short form": "Possess nothing, own none, / Nakedly come, nakedly go" (Wang, 109).

Visually, Wang's woodcut presents a rollicking primeval whorl of English words, Chinese figures, and erotic images. His truncated couplet intertwines block letters and thorny serifs with iconic vines and twigs. Nude Venus figures, black and white, lounge in sun and shadow, hands clutching lush clumps of untamed vegetation, eyes cut to resemble wildflower sunbursts, effluvial hair flowing like rivers of rain. Pubic hair and thistles bristle. Penis-shaped stalks appear poised to inseminate a supine female figure sprawled in ecstasy. In the center, a sandpiper-like waterbird, set in an oval, resembles the circled figure of a naked female torso whose vaginal beak and eye-like breasts mirror the image of the bird. In substance and form, Wang's woodcut communicates the extent to which Francis, in his treatment of the rural erotic, explores through a male-oriented field of vision the metaphysical and epistemological weld between sex, gender, and the natural world.

Generally, Francis's literary and cultural focus constitutes what Alan Sinfield refers to as an exploration of "non-metropolitan sex-gender systems" and what Richard Phillips calls "sexualities in non-metropolitan contexts" (Sinfield, 28; Phillips and Watt, 9).[1] In some ways, the gender

border crossings in *Gusto, Thy Name Was Mrs. Hopkins* serve as preludes
to the eco-erotics in Francis's experimental magnum opus in miniature,
A Certain Distance. In 1976, *A Certain Distance* culminated Francis's
lifelong interest in celebrating the rural and the erotic as a vicarious iden-
tification and self-study, in the same sense that Whitman's multi-singular
song of the "self" encompasses innumerable human and "more-than-
human" worlds, as J. Scott Bryson and Michael McGinnis have termed
them (Bryson, *Ecopoetry*, 8; McGinnis "A Rehearsal" 8). Even before *A
Certain Distance*, however, Francis was sowing the seeds of this hybrid
aesthetic in the cultural furrows of his twentieth-century New England
bioregion, which in some instances extended no farther than Amherst's
Market Hill Road or the backwoods behind his house.

Throughout Francis's writing career, he composed works that occu-
pied and altered the marginal spaces created by the fusion of *eros* and
bios.[2] For example, in "Two Uses," the speaker sees the twofold giving-
and-receiving capacity of the human body's erotic dimensions as a sign
of "nature's thrift" (7). Every so often, Francis's rural musings strike
readers as lighthearted, other times, as seriously meditative and raptur-
ous. When Francis, in the spring before he died, read "Alma Natura" on
Henry Lyman's *Poems to a Listener* radio program, he prefaced his read-
ing by saying that people sometimes saw him as a "mischievous" poet.
The speaker in "Alma Natura" unabashedly ogles Mother Nature as a
pin-up girl, praising her ample "bust no bra could hope to capture" and
wishing to see her more than barefoot: "Ah, more than barefoot, more
the better" (6–7). In "Comedian Body," a poem that Donald Hall called
"amused and lonely at the same time," Francis's blithe couplets celebrate
the "bawdy" nature of the male physique, the "poor fanny / So basic and
so funny" and the "penis pun / That perfect two-in-one" (Hall, "Two
Poets," 123; 2–4). In "Floruit," however, the joking yokel turns transcen-
dent Yankee lover and melds images of the erotic and the geocentric in
two elegant one-sentence stanza-streams of paradoxically chaste ravish-
ment. Similar in theme to Donne's devoutly sexual submission to the will
of a three-personed God, "Floruit" treats sexual satiation as part of the
grander ecosystemic process. "Daringly, yet how unerringly," the speaker
opens, are all the "cool and nun-like virtues" of wildflowers in the pres-
ence of all the "hot virtues of the sun / And being wholly sex are wholly
pure" (1, 5–6). Having established the natural world's superior and pure
erotic essence, the speaker challenges the human race: "If with an equal
candor we could face / Their unguarded faces, if we could look in silence
/ Long enough, could we touch finally, / We who when luckiest are said
to flower, / Their fiery innocence, their day-long unabashed / Fulfillment,
the unregretful falling?" (7–12). Simply put, in "Floruit" nature schools
humankind in the nature of erotic love.

Previous to the fulfillment he would achieve in *A Certain Distance*, Francis, as naturalist-philosopher of the erotic, authored many shorter pieces in which the relationship between sexual desire and biospheric phenomena creates a balanced reciprocity within David Greenberg's "egalitarian" erotics (Sinfield 27). We might also class Francis's early- and mid-twentieth century ecolyrics within the ranks of the modernist "gay pastoral" that David Shuttleton highlights.[3] Though Francis's forms vary, his aim remains centered on embracing and preserving in art his erotic and ecological orientations. In the coupled acts of reading and writing, the erotic and ecological gravitate toward becoming one and the same. Since "Francis feared to touch the objects of his desire," Alan Sullivan writes, "his principal sexual organ was the eye, which dwelled lovingly on the male form in pose after pose, poem after poem" (Sullivan, par. 12). What Sullivan observes is true—that to Francis, the erotic and the rural operate a reflective dialectic, mirroring and mutually transforming each other.

The nature prose/poetry collected in Francis's *A Certain Distance* evolved over nearly a half-century. Of the thirty-one stanzaic and paragraph-type compositions, fourteen are reprinted. This means that nearly half of *A Certain Distance* derives from previous publications: one piece from *Stand with Me Here* (1936), six from *The Orb Weaver* (1960), five from *Come out into the Sun* (1965), and two from *Like Ghosts of Eagles* (1974). This breakdown indicates that while Francis postponed his most subtle and daring literary exploration of the rural erotic until 1976, he was forming its foundations forty years earlier. In a certain sense, *A Certain Distance* serves as the twentieth century's miniature *Leaves of Grass*, though its volume constitutes a fraction of the space filled by Whitman's *meisterwerk*. Even the title of Francis's consummate work-in-progress underwent significant alteration. In the Tetreault and Sewalk-Karcher interview, he mentions "two continuing projects," one being a "collection of portraits and sketches" called *Observations and Visions of the Young Male* (*Pot Shots* 133). This openly homoerotic title became the more ambiguous *A Certain Distance* for unknown reasons, perhaps in response to more conservative publisher demands. And yet, as a title, it captures the paradoxical combination of fixity and openness central to Francis's view of the erotic dimensions of the ongoing interrelationships between human and more-than-human natures. In a word, "certain" connotes definiteness as well as lack of definition. We may be certain of our views about sexuality, but we may also walk a certain indeterminate distance down a wooded road toward a general destination. It is this certainty and lack of certainty concerning elements of the rural homoerotic that Francis takes for his focus in the nature prose/poetry in *A Certain Distance*.

Francis's fiction and early poetry divided critics and readers, and his later experimental nature prose/poetry delivered more of the same—questions rather than definitive answers.[4] Though some saw Francis's prose-poetic meditations as evidence of a repressed and sexually frustrated psyche, more accurately, *A Certain Distance*'s "contemplations and celebrations of young men," as Howard Nelson calls them, employ a "looser" form to communicate a transcendent love, an aesthetic synthesis of the homoerotic and ecological whose *via affirmativa* replaces bitterness and aggravation with beauty and placid observation still supercharged with passion (Nelson, 10). In his introduction Francis clearly stakes out his territory. His dedication of his work to Michelangelo and the Sistine Chapel's twenty-six Adam frescoes conveys the lofty seriousness with which he undertakes his project: "I wrote—I drew, I painted—these pictures for the pleasure of doing them. . . . Many an artist over the centuries has concentrated on the female form. . . . These pictures of the young male can be accounted for, I suppose, by the same psychology" (*Certain Distance*, 1). A thorough reading of *A Certain Distance* reveals a loosely arranged but orchestrated methodology that places the pieces in three categories: Euro-Italian, reprinted, and local.

We can trace Francis's Euro-Italian sketches in *A Certain Distance* to his experiences abroad. The year before *A Certain Distance* was published, Charles Sides observed, "Perhaps conventional attitudes toward homosexuality explain why [Francis] has written no direct love poems" (Sides, "Freedom," 70). No doubt readers who seek direct, explicit treatment of the male *eros* in *A Certain Distance* will be disappointed. As a literary text, its strength lies in its subtlety, its "Grace of Indirection," as Robert Bly labels Francis's style (Bly, 101). In the Euro-Italian sketches, the male *eros* is hinted at, glancingly referred to, and teased out of the most peripheral detail. These sketches reveal as much as they conceal and achieve amplified titillation through deft diminishment.

A clutch of Euro-Italian pieces eroticizes the everyday, the momentary, and the unexpected. "I don't know what I admire about him most," Francis begins "Mechanic," ending decisively with "the way he comes up close to me when he has some explanation to make and looks me straight in the eye" (*Certain Distance*, 6). Here, a lack of distance and the mere act of looking constitute a greater erotic exchange than touching. In "Young Frenchman," which Rudy Kikel labels "sexually ambiguous," the speaker observes the erotic male subject at a dance and makes verbal contact: "At last, coming face to face with him at a doorway, I spoke. The change in him was dramatic. Instantly, he was all attention, all alertness" (Kikel; *Certain Distance*, 38). In an exchange of homoerotic energies, the speaker describes the power he possesses, through observation and linguistic composition, to transform himself and the subject of his interest: "Picture him on a deck, dressed in sandals, blue shorts, and one of his sister's straw hats as a shield against the sun. A spirit perfectly balanced, perfectly free" (*Certain Distance,* 22–23). In "Young Man of Assisi," Francis's homoerotic subject becomes the young "filling station attendant" in the Piazza Santa Chiara. The speaker confesses a fascination for the attendant, who "unlike other young men" projects a look of "serenity and sweetness" and who while comparable to a "monk" shares a secret love relationship—with the speaker: "I concluded that he must have been in love and that the love so filled his mind there was room for nothing else. Weeks later in Rome with all its confusion and distraction I kept thinking of the young man of Assisi as I walked along the street" (*Certain Distance,* 20). Here the urban environment of Rome enhances the homoerotic gravity between the speaker and the beloved. Francis's Euro-Italian sketches, though more urban than rural, operate via the notion of "embodiment" as Daniel Spencer explains: "[E]mbodiment includes sexuality, but is a broader recognition that everything we do and all that we are is mediated by our bodies. Embodiment in an ecological paradigm reflects our relation to each other and the rest of the earth, as well as our access to the good of the earth such as healthful nutrition, labor, and rest" (Spencer, 107). As embodied figures of homoerotic and environmental significance, Francis's male subjects fuse love of man and love of the earth's land and cityscapes.

These portraits also convey a certain penchant for working-class erotica that readers discover in the reprinted pieces as well as those more rural and local in content. Francis's poem "Bronze" appeared in his 1936 collection, *Stand with Me Here,* and later in *A Certain Distance* and his published journals. In this case, Francis appears to have given long life to a brief poem: "Boy over water, / Boy waiting to plunge / Into still water / Among white clouds / That will shatter / Into bright foam—/ I could wish you / Forever bronze / And the blue water / Never broken" (1–10). Francis's affinity for "bronzing" his male subjects carries a twofold significance. In one sense, it figuratively covers the male subject in a metal associated with endurance, outdoor health, vitality, vigor, and strength. In addition, it recasts the erotic subject as a statue or work of art. As an earth-borne element, then, bronze fuses the natural, aesthetic, and homoerotic. In his journals (July 31, 1932), Francis details the process of composing "Bronze," a poem, he says, that was in his mind "for at least two years." He notes, "The writing of poetry being my most deeply creative act, I feel toward the slightest marring of the atmosphere, the slightest hint of interference not only with my activity but with my thought and mood, as one feels during sexual orgasm. Interruption is simply intolerable" (*Travelling in Amherst,* 15). In this passage, sensitive to the smallest atmospheric disturbance and aware of the two-year fertilization period required to grow a single ten-line poem, Francis articulates a unique insight concerning the interrelationship between nature and eroticism through literary composition. In nature and writing, he suggests, little, if anything, divides the organic and the orgasmic.

The reprinted compositions in *A Certain Distance* suggest that Francis's end-of-life project was a lifelong endeavor. The collection contains six pieces of rural homoerotica reprinted from *The Orb Weaver*, which Francis originally titled *With the Year's Cooling (Frost: A Time,* 11). In the seven free-verse couplets of "The Rock Climbers," the speaker, from the perspective of "soft / middle age," finds his deflated anima electrified by the climbers who "giving all to love / embrace cold cliffs / Or with spread-eagle arms / enact a crucifixion / Hanging between the falling / and the not-attaining / Observed or unobserved / by hawks and vultures—" (3–10). Though he sidesteps the connection between the geocentric and homoerotic in "The Rock Climbers," Mark McCloskey notes the poem's paradoxical tension: "Man must . . . strain but fail to achieve his aims, and experience the elation and despair of recognizing this truth. Moreover man is a construct of belief and skepticism, freedom and restriction, communion and loneliness, for which polarities he is both praised and damned, distinguished and forgotten" (McCloskey, 272). This strain depicts erotic desire against the backdrop of the earth's firm crags and majestic raptors—all representative of the passion that the aging male lover desires but cannot possess or experience alone. Paradoxically, the sinewy climbers serve to inflame the longing and cool the desire of the elder spectator: "How vaulting a humility / superb a supererogation / Craggy to break the mind / on and to cool the mind" (11–14). It is as if, for Francis, to reprint his work meant to revisit the mental and physical erotic charge of youth through the mind and soul of an older man.

Many shorter reprinted pieces celebrate the simple physicality and energy of male youths, energy that is both kinetic and static. "Watching Gymnasts"—dedicated to a then-young Paul Theroux—praises the "flower-light" movements of gymnasts (9). In "Metal and Mettle," a quotidian spectacle of working-class physical beauty attracts the speaker, who confesses his inability to "choose between" the voyeuristic "esthetic satisfaction of watching" either the "slow laborer" or "speedy athlete" (1–3). Though the athlete presents a stunning figure, Francis's speaker appears magnetically drawn toward the laborer whose "bronze back is pushing / a wheelbarrow level with cement up / an inclined gangplank steadily / the center line of the back undulating / at each step like a cobra delicately / dancing to a flute" (3–8). "Farm Boy after Summer" meditates on the "bronze slowness" that "becomes" the motionless farm boy who seems a "seated statue of himself" and whose "mind holds summer as his skin holds sun" (1–2). "The Base Stealer," an often reprinted and athletic *tour de force* of linguistic agility, describes youthful erotic fulfillment in terms of a rural baseball game: the boy figure "tingles, teases," and "[t]aunts" the pitcher and his lover like an "ecstatic bird" (7–9). "He's only flirting," the speaker says, piling nine lines of arousing description up before releasing readers with the climax: "Delicate, delicate, delicate, delicate—now!" (10).

The balanced two-part, two-stanza poem "Swimmer" constitutes, as David Graham observes, one of Francis's "happier poems" that possesses a "definite undercurrent of harshness" (Graham, 91). Deliberately tranquil in its treatment of the rural homoerotic, "Swimmer" describes the quest for environmental and sexual wholeness in terms of a Hegelian dialectical struggle: "Observe how he negotiates his way / With trust and the least violence, making / The stranger friend, the enemy ally," the speaker begins, deliberately infusing the words "stranger," "friend," "enemy," and "ally" with ambiguity so they refer to both water and the lover (1–3). "The depth that could destroy gently supports him," he observes, adding, "Danger he leans on, rests in" (5–6). In Part II, the erotic elements of the poem enjoy greater explicitness. "What lover ever lay more mutually / With his beloved, his always-reaching arms / Stroking in smooth and powerful caresses?" the speaker asks (8–10). Characteristically, Francis demonstrates how a poem functions as an open-ended equation. "Some drown in love as in dark water, and some / By love are strongly held as the green sea / Now holds the swimmer," he writes, finishing, "The swimmer floats, the lover sleeps" (11–14). David Graham highlights the "multifoliate meanings" in "Swimmer" and argues that it compares "languid lovemaking to swimming" and self-realization as obtained in the "dangerous water of otherness" (Graham, 91).[5] Francis further develops this brand of rural homoeroticism in reprinted pieces from the 1960s and 1970s.

On one hand, select compositions in this category appear to unfold with more verbal and physical immediacy, as if the speaker records what someone has said or done. In "When I Come," the speaker mingles stark environmental tableaux with snatches of remembered dialogue uttered by a male counterpart. The substance communicates a deep and abiding longing between the two men. The language and style remind readers of Ezra Pound, especially in the final stanza:

Once more the old year peters out—
all brightness is remembered
brightness.

> (*When I come, Bob,*
> *it won't be while just on my way*
> *to going somewhere else.*)

A small pine bough with nothing
better to do fingers
a windowpane.

(When I come, Bob—)

Against the wet black glass a single
oval leaf fixed
like a face.

Similarly, "His Running My Running" departs from conventional gram-
matical and mechanical strictures and through compact "epigrammatic
force," as Victor Howes describes it, meditates on a more autumnal eroti-
cism, both in terms of the age of the human speaker and the age of the
earth (Howes). In this poem, the speaker, during a time he identifies as
"mid-autumn late autumn / At dayfall in leaf-fall," vicariously inhabits
the physical experience of the youthful runner by writing, "His lone-
ness my loneness / His running my running" (1, 14–15).[6] Donald Hall
sees an erotic and environmental identification in "His Running My Run-
ning" wherein the "eye of the poem . . . returns to the observant present
by feeling itself into the runner outside." Further, Hall notes, "There
is something attractive about those 'bare legs gleaming'; the watcher's
furtiveness comments on the attractiveness" (Hall, "On 'His Running'"
312). Though at times Francis treats the rural erotic with a universality
that spans cultures and topographical boundaries, he exhibits the ability
to approach the subject in a setting closer to home.

In *A Certain Distance*, erotic does not necessarily mean exotic.
When Victor Howes classed Francis as a "local poet who looks out on an
expanded horizon," he appeared to prophesy the substantial number of
local erotic pieces Francis would generate (Howes). Noticeably, several of
the local pieces treat intellectualism as a source of homoerotic attraction.
In "Two Phases," the speaker's overnight guest undresses "to his under-
wear," reclines on the speaker's sofa that "had been made up as his bed,"
and goes on reading a "book he really wanted to read." In the morning,
the speaker looks in on his guest, admiring the slumbering figure "in the
full bloom of manhood" who resembles "Ulysses," "Holofernes," "or John
the Baptist on a salver, painted by, say, Caravaggio" (*Certain Distance*,
34). In "Philosopher," another guest reads at the speaker's home, and
the speaker catalogues his guest's physical appeal: "lips, slightly parted"
and "dark visionary eyes and intense hands!" (*Certain Distance*, 13).
Another male guest uses the speaker's shower in "Heat Wave," deluging
his body in hot then cold water that transforms his hair into "moss on a
mossy stone in a waterfall." "Still it pours and pounds," the speaker mar-
vels, from a voyeuristic vantage point, "The insatiability of it" (*Certain*

Distance, 15). "In Control" describes the way a "handsome" and "dash-ing" local man of "confidence" crosses the street: "He moves as a man should move, briskly, easily, his whole body a unit" (*Certain Distance,* 16). "Man in a Dark Red Jersey" catches a "lean, deeply tanned chap" of "mostly bone and muscle with just enough flesh over them" in the act of unbuckling his belt and tucking in his shirt, whose "slow, easy, rhyth-mic swinging movement from the shoulders and hips" leads the speaker to conclude he is a "young construction worker" or "student who had done much manual labor" (*Certain* 30). In his autobiography, Francis includes passages that describe his twin love for rurality and working-class individuals as a ripe combination for artistic inspiration. In an *ars poetica* invocation, he exults how in his early days as a rural writer out-side Amherst, he was "glad" that "there wasn't a single college graduate . . . in the whole area known as Cushman, or anyone who might have qualified as an intellectual." He remembers that he felt "more comfort-able among plain people—laborers, janitors, factory workers, and small farmers" than those in a "college community" and found "such people more stimulating to [his] imagination" (*Trouble with Francis,* 14). Space and a rural place to contemplate the allure of the male form were essen-tials to Francis's mode of living and writing. In some ways, he could not write about one without the other.

Overall, the prose-poetic sketches in Francis's *A Certain Distance* invite readers to see, as Daniel Spencer has written, "the erotic as ecolog-ical and ecology as erotic" (Spencer, 321). In the same way that Francis's rural sketches, portraits, and lyrics locate the orgasmic in the organic, they seek to map the certain and sometimes uncertain distance between the erotic and biotic.

Male Venus, Eco Homo

It is problematic at best to derive biographical and scientific "facts" from an author's life based on the writing he produced. Likewise, we can-not necessarily tie "truths" about what a work of literature "means" to the author's personal history and biology. In Francis's case, however, select writers and critics have attempted to do just that. For example, Elinor Phillips-Cubbage classes Francis as "a man with sexual hang-ups (i. e. the Puritan conscience)" (Phillips-Cubbage, 2, 139). Alan Sullivan, too, claims that the practice of Zen served as an "antidote" for Francis's "sexual frustration," elsewhere calling Francis a "boy-smitten old queen" (Sullivan, pars. 34, 38). "An aging closet queen" is the label Rudy Kikel suggests some gay readers may hang on Francis (Kikel). However, these conclusions tend toward "reductionist" models, as Daniel Spencer calls them in *Gay and Gaia*. Citing the work of Donna Haraway, Spencer cautions against overly dualistic and essentialist cultural paradigms and outlines the difficulty of classifying sex, gender, and nature in terms of either "scientific realism (the position that what science portrays is objectively true)" or "social constructionism (what science portrays is merely a reflection of the dominant societal power structure)" (Spencer, 86).

For decades, scholars have discoursed ably on the subject of sex as biologically and scientifically determined and gender as naturally, socially, and culturally constructed. It has not been my aim to assign any such finalized definitions to Francis, nor to file him in any narrow cultural compartment. As an innovative nature writer who challenged the limits of genre and as a human being who inhabited and altered the boundaries that tend to define genders and non-human habitats, he still defies categorization. Pushed for a label, we might think of Francis as one of Michael McGinnis's "boundary creatures." McGinnis sees every human and animal as a boundary creature that "inhabits more than one world." "The salmon, bear, and people," McGinnis adds, "are linked and are nested 'parts' of several distinct but interdependent systems of relationships" (McGinnis, 4). As a boundary creature influenced by gender-inflected relationships and an ongoing fascination with and absorption in the earth's ecosystemic processes, Francis became a unique breed of writer, perhaps one of a kind. His beginnings, thanks to his writing, are by no means mysterious. More authoritatively than any scholarly or theoretical opinion, his words communicate a sense of how he became the person he was.

In his autobiography, Francis details many experiences that shaped his erotic orientation and sense of gender constitution—including early recollections of playing with a doll named "Bessie Bun" (which his mother later removed from his room without telling him) and the memory of youthful and innocent "erotic gropings" with a small neighbor girl named Catherine, an incident for which he was "found out" and "denounced and disgraced" by the girl's mother (*Trouble with Francis*, 140–141). In chapter seventeen, "Eros," Francis records his adult homosexual relationship with an unnamed Italian man to whom he attributes "high spirits" and "great gusto" (*Trouble* 212). While Francis mentions but does not name his Italian love interest in print, collected correspondence and other surviving archival materials suggest that Francis may have invented this male Italian paramour as a cover for another individual.

As for the experience itself, though, Francis felt he had nothing to hide. "Far from being ashamed of it," he writes of that time in his life, "I never ceased to give thanks that I had at last reached one of the great experiences of my life" (*Trouble with Francis*, 210). His account of this dimension of his life is memorable, given its carefully wrought composition and

sublime reflectivity. While he openly discusses his lifelong erotic attrac-
tion to "members of [his] own sex and to them only," he declares that
he was "unwilling to accept eros or even to look for it without tender-
ness and mutual respect" and states he "preferred to go without, and was
prepared to go without." Most captivating is his sense of inner harmony
and balance: "What I was above all determined to have was a good life
in which no single element however urgent would dominate and distort
the rest. Though eros might pervade my thought, it did not usurp control
over my actions" (*Trouble with Francis*, 211). Notable here is Francis's
resolution that erotic love not dominate his life but instead serve an inte-
grative function within the greater systemic interrelationships formed
among his life, others' lives, and the life of the earth. Elsewhere Francis
writes about periods in his adult life when this awareness and orientation
reached fruition.

In chapter 6, "Peace at Fort Juniper," which covers the period from
1945 to 1954, he recounts his sunbathing and rain-bathing practices. At
the same time, he demonstrates how he had through the course of his
solitary existence in the wilds outside Amherst come to live in a way that
celebrated an embodied experiential fusion of the erotic and the biotic,
particularly during his frequent nude sun baths: "What made the lux-
ury all the greater was the completeness of my nakedness. Others might
enjoy a ninety-five percent nakedness on some beach; my enjoyment was
one hundred percent. For me it was a kind of love affair, the sun being
my magnificently powerful and infinitely gentle lover" (*Trouble* 55–56).
Clearly, the rural erotic was something Francis not only wrote about but
lived. In a very real sense, the erotic component of his natural environ-
ment satisfied the need for fulfillment he chose not to find elsewhere in
the human world.

If Francis is not a backwoods nature boy confined to the small-print
footnotes in American environmentalist history, how are his life and art
still relevant? David Bell points out that many "commentaries on nudism
. . . stress the decoupling of nudity from sexuality, emphasizing a spiritual
chaste purity" (Bell, 83). Further, though, Bell argues that "naturism is
profoundly about the erotic" and cites Greta Gaard's *Toward a Queer
Ecofeminism* and its focus on the "sexing of nature and the naturing of
sex" (Bell, 85, 93). Bell's exploration of "naturism" and "sexualizing the
country" describes Francis's lifestyle and literary practices. Together,
Francis's mode of living and literary output invite readers to reconsider
the ways in which the erotic and the rural construct and modify one
another. Further, Bell, citing the work of Mark Lawrence, highlights the
erotic as a "given sign" of the rural, adding that as a "metalanguage of
nature" it serves as a "constellation of made, unmade, and remade con-
structions." "At the same time," Bell argues, "*the rural*, as a given sign of

the erotic, also evokes a constellation of meanings" (Bell, 97). Constellations, by nature, are visible only because many points of light shine in unison. It is precisely a cultural lack of visibility, or "invisibility and silence," concerning lesbians and gay men that Daniel Spencer argues "typif[ies] many ecological models" (Spencer, 63). During Francis's lifetime, his light pulsed and burned steadily, though not brightly on a national or global scale. Surely, his singular literary life deserves a more visible place in the growing constellation of American ecopoets.

In the Francis issue of *Painted Bride Quarterly*, Fran Quinn writes, "[Francis] was gay although he acted so seldom on his sexual interests 'asexual' often seems more appropriate" (Quinn, 6). Given Francis's status as solitary brooder, observer, recluse, and nature writer, it is difficult not to think of him as beyond definition, as a local wayfarer, minor holy man, quiet cultural warrior, Transcendental anachronism, aesthetic dissident, or fringe philosopher. It may also not be too far afield to accept him as the author of the "third genre," as well as an author representing the "third gender." His lifestyle and work suggest that maleness and femaleness, mediated and absorbed by the third gender of nature, begin where the other leaves off, just as "poetry may begin where prose leaves off," as he quips in his miniature essay "Silent Poetry" (*Pot Shots* 61). "In Native American cultures, in Japan and elsewhere," Alan Sullivan reminds us, "a third sex was recognized" (Sullivan, par. 19). Also, in examining Native American cultures, Daniel Spencer highlights the spiritually and ecologically oriented social function of the "berdache" or "alternative 'third gender.'" Spencer writes, "The berdache . . . is a morphological male with an androgynous, non-masculine character who often held a place of esteem and respect" (Spencer, 335). Referring to the work of Michael Clark, Spencer spotlights the function of the "ecological berdache" as pursuing the goal of "the joint valuing of human- and biodiversity, integrating the spiritual and the physical, the erotic and the ecological, to engage in healing and restoring *all* the earth" (337). While the identification of Native American cultures with white Americans is undoubtedly "problematical," as Bernard Quetchenbach points out, and while Francis does not qualify as an ecological berdache in terms of indigenous ethnicity, his status as erotic, geographic, and economic outsider would cast him as a literary, cultural, and ecological revisionary (Quetchenbach, 22).

Today Francis's legacy may be that his life and writing do not ask us to behold poem or prose, human nature or non-human nature, woman or man—but simply to behold.

4
Fiction and Non-Fiction

THROUGHOUT HIS WRITING CAREER, FRANCIS appears to have been torn between identifying himself as either a poet or a writer of prose. While he claims in his journals that the poet in him eventually triumphed over the informal essayist, the nature and scope of his total output would indicate that he excelled as an artisan of both stanza and paragraph. The unique species of American ecopoetics that Francis engineered is flexible and capacious enough to encompass his prose writings, both fiction and non-fiction, published and unpublished. His one published novel, *We Fly Away*, can be read as an example of what ecocritics have called the "bioregional narrative," a synthesis of wild nature and human culture, a call for human and non-human communities to be linked in mutual sustainability, compatibility, and reciprocity. In addition to this longer work, the nearly two hundred informal essays that Francis published in *Forum* and *The Christian Science Monitor* capture the multifoliate symbiosis he sought between human and non-human life forms.

FLIGHT

In 1984, James Merrill took the podium in front of an assembly at the New York City Library. The audience he addressed had gathered to witness Robert Francis receive the honorary title of Life Fellow in the Academy of American Poets, a distinction that carried a $10,000 award. While Merrill delivered primarily ceremonial remarks, he also used the occasion to request that Francis's work be "more widely read" and called for someone to reprint Francis's "beautiful novella, *We Fly Away*" (Gillman, "The Man"). Over twenty years previous to his public praise, Merrill privately expressed similar feelings about Francis's only published work of fiction, calling it a "lovely, moving experience" (Merrill). Though neither Merrill nor Francis—nor anyone at the New York City Library, nor around the globe, for that matter—would have likely referred to *We Fly Away* as a bioregional narrative, that is exactly what Francis gave the world: a lucid tale of a quiet radical who sheds the strictures of social convention in order to live in greater harmony with his environment and natural surroundings.[1]

We Fly Away, a bioregional narrative now all but buried beneath the shrapnel and sediment of post-World War II American fiction, divided those who read it. Some readers found themselves enlightened and invigorated; others merely yawned.[2] The mixed reaction that *We Fly Away* received may have been due to its maverick spirit, serene content, simple style, and baffling structure. Even today, as a novel (or novella), it defies convenient categorization. In some ways a thinly veiled autobiography, *We Fly Away* reads as if poised ahead of and behind its time, an earth-sensitive, social-activist tale that chronicles the birth of the artist into the wild world, an environmentally tuned *Künstlerroman*. As brief as the breeze, it occupies an indeterminate slot between realism and sentimentalism, subtlety and blandness, admirable Thoreauvian simplicity and problematic oversimplification. In response to Merrill's charge, I offer a twenty-first century reexcavation of *We Fly Away* in terms of its substance, structure, and style.

In the area of substance, three interlaced stories, like fibrous veins of bark on a limb, comprise the narrative in *We Fly Away*: the story of Robert, Mrs. Bemis, and Mrs. Teal (the domestic); the story of Robert and the young student boarder, Henry (the homoerotic); and the overarching story of Robert's transcendent absorption and artistic assimilation into the surrounding wilderness (the eco-aesthetic). In terms of conflict, Francis sets his protagonist, the thirty-five-year old aspiring poet named Robert, against his landlady, the "slightly acid" Mrs. Bemis (anon. review). Mrs. Bemis represents the paralysis of domesticity, the crotchety curmudgeon, the elder middle-class woman who has lapsed into a folksy, blind insensitivity toward nature. To contrast her to Robert, Francis describes Mrs. Bemis, who rarely leaves her creaky sarcophagus of a home, in terms of religious ritual, death, and kitsch art—as an "acolyte," in "black silk," as "motionless as a Currier & Ives" (*We Fly*, 26, 35, 47). Early on in the sedate action, Robert, on returning from a lecture at the local library, pauses outside to watch through the window as Mrs. Bemis reads: "He waited to see her turn a page. But she did not turn a page. She made not the slightest motion. If she were living, as she doubtless was, she was as still as death. If she were dead, as she probably was not, she was as real as life, . . . with painted background in an illuminated glass case" (*We Fly*, 26). Later, Robert muses on Mrs. Bemis's eventual demise, the image of her prone, lifeless body on her bed: "Yet having been at home with death so long and still living, she would seem eternally at home with death and never dead" (*We Fly*, 31). Throughout the narrative, death shrouds Mrs. Bemis like a cloud.

As a character, Mrs. Bemis evokes a sense of superficiality, artificiality, unnatural containment, and separation from nature. She also functions as the bungling technophobe whose egocentric attempts to master rather than be mastered by nature strike readers as comically satirical. Mrs. Bemis claims that a lawnmower nearly runs her over in her garage, expresses her desire for snow to fall no more than "two inches at a time," scolds Mrs. Teal for catching a cold, agonizes over the premature bloom of her crocuses, and then abruptly alters her position: "It was the season itself that was backward" (*We Fly*, 99, 132). Mrs. Bemis's egocentric domesticity conflicts with Robert's progressive ecocentric retreat into unmediated interaction with his natural surroundings. At times, Mrs. Bemis, as Robert's counterpart, serves to draw him out, helps develop him as her opposite, and unconsciously bequeaths the philosopher nature-poet his ideological claws. As a semi-resistant provider of Mrs. Bemis's household help, Robert conscientiously classifies himself as a "disbeliever in rugs on aesthetic, sanitary, and economic grounds," which echoes *Walden* when Thoreau resists the infiltration of carpets and mats as "the beginnings of evil" (Thoreau, par. 2). Moreover, Robert's conservationist stance lends him newfound wit. When Mrs. Bemis incinerates a housefly in her oven, Robert protests in favor of the fly's right to live, to which Mrs. Bemis condescends that flies exist merely to "discipline" humans. Robert's gnomic reply buzzes beyond Mrs. Bemis's ken: " 'I see,' said Robert. 'Flies discipline us and then we turn around and discipline them. I think we do a better job than they do.'" Later, when Mrs. Bemis rankles at why New England winters must be so cold, Robert echoes wittily, "To discipline us, Mrs. Bemis" (*We Fly*, 62, 107). Overall, Francis depicts Mrs. Bemis as "beautifully predictable," and when the budding naturalist-poet Robert observes that he, too, is "becoming predictable" and "part of the pattern, a part of the picture," readers sense it won't be long before the caged, fledgling artist will fly to the hills to roost and ramble with his vagrant soul (*We Fly*, 20).

Mrs. Teal, Mrs. Bemis's "fretting housekeeper" and cook, functions as deuteragonist and quintessential quidnunc (anon. review). She acts as inspirational precursor to Robert's eventual "flight" from the daily doldrums of Mrs. Bemis's cooped-up routine when, after a brief stay at a "Rest Home" due to illness, she makes the decision to leave Mrs. Bemis's employment for a more profitable, rival housekeeping position under Mrs. GlenEllis (*We Fly*, 94, 121). At first, Mrs. Teal asserts her superiority in the Bemis house's pecking order, insisting that Robert restrict his furnace-tending duties to ash removal while keeping the remaining furnace responsibilities for herself. Later, however, after Mrs. Teal's irritation erupts and, in silence behind her unsuspecting employer, "like a wraith . . . with lifted brows and clenched teeth she [shakes] her arms, as an avenging fury," Robert notices that Mrs. Teal begins to sneak him extra muffins from the kitchen and that, prior to her leaving, "a change had come over Mrs. Teal" (*We Fly*, 63, 118). At the novel's conclusion, Mrs. Teal and Robert languish in a twin state of domestic disharmony, supersaturated by over-civilization and unacquainted with the sense of

reciprocity, compatibility, and balanced community the bioregional nar-
rative evokes. On the wings of Mrs. Teal's departure, however, Robert
clears virgin existential terra firma and transcends imminence so that his
outdoor surroundings become a central "player" in the community of two
he gradually composes between himself and nature.

Henry, Mrs. Bemis's student boarder, functions as "Robert's exact oppo-
site," a "synthetic and symbolic" character, Robert's homoerotic interest
and poetic muse (anon. review; F. P. R.). Through Henry, the struggling
poet Robert discovers a reawakened sense of desire and new direction,
which feeds and fertilizes Robert's hunger for art and nature. Henry's
entrance into and departure from the novel, both somewhat mysterious
(and perhaps forced and arbitrary), mimic the migratory arrival, flight,
and demise of a predatory raptor. When Henry arrives—perhaps from
Pennsylvania or West Virginia—to room in Mrs. Bemis's house, Robert
thinks "pleasantly of the prospect of a fellowman in the house." Later,
he describes Henry as "a substantially built youngster with something
of the sleekness and glint of a well-groomed colt." Robert also praises
Henry's cough as "a good resounding masculine bark" (*We Fly*, 19, 23).
Francis's depictions of Henry, through the eyes of admiring protagonist
Robert, fuse the homoerotic and aesthetic: "He strolled into the dining
room, picked up the receiver, and made his call with such casualness,
such complete lack of self-consciousness, that the whole episode had the
style and finish of a work of art. Once or twice he laughed softly over the
telephone. The voice was pitched low and resonant" (*We Fly*, 110). In a
short episode that parodies the *Romeo and Juliet* balcony scene, Robert
lets Henry in at night after Henry forgets his house key. "Let's go in my
bedroom," Robert suggests to Henry, looking at the young male student
with a "mixture of admiration and doubt"—an invitation Henry casually
declines (*We Fly*, 30). At times, protagonist-poet Robert appears to craft
his relationships with Henry as if carefully revising the draft of a poem.

But the attraction remains one-sided. As erotic subject, Henry jokes
about Mrs. Bemis's antiquated use of the word "rubbers," and later, after
Henry gets drunk and Robert helps him into bed—divesting the vul-
nerable young man of shoes, shirt, and pants—the intimate experience,
rife with sublimated passion, produces an impulse that fuses the homo-
erotic, natural, and artistic: "The trees, the roofs, and the ground all lay
asleep, having drunk the moon's golden liquor. Henry in his bed, and
the old house on its foundation were in the same state tonight. Strangely,
Robert still seemed to smell the alcohol of Henry's breath. But more
strangely, the whole episode had ceased to be a disaster and a disgrace
and had become a sort of ritual, very old, very earthy, full of humor and
poetry" (*We Fly*, 75–76, 82). This primeval tuning to the earth's rhythms,
linked to desire for Henry (whose very breath intoxicates Robert), acts

as combustive muse and inspires Robert to rummage through his work-in-progress poetry manuscript—to revise a poem, "By Night"—in order to see his writing through "another person's eyes" (*We Fly,* 76). As the plot advances, the two men gravitate physically together while remaining emotionally apart. Previous to Henry's accidental death, which one writer erroneously labels "incidental," Robert agonizes self-consciously over inviting Henry to share a cup of coffee or cider in the kitchen, and Henry insouciantly invites Robert to come with him to Florida (*We Fly,* 111, 113). Symbolically, during the late February chill, Henry travels south to Florida, and Robert, inspired by his migratory muse, sends his poetry manuscript south, too, to a publisher, "though only far as New York." And when Robert reads in the newspaper of Henry's death in a motorcycle crash, he blinks "sudden tears out of his eyes" (*We Fly,* 114, 153). While Henry's inclusion in the plot smacks of stiff convenience— and his name functions as a somewhat facile allusion to Thoreau—his character operates as poetic muse, homoerotic love interest, and worn pathway to the natural artistry outside Robert's provincial prison.

If anything, *We Fly Away* remains a bioregional narrative about one phenomenon: vision. In many ways, it constitutes a re-visionary text, an invitation to turn one's sight outward, from egocentrism to ecocentrism, for both Robert and the reader. The main narrative vein that fuses all the others involves Robert's hypnotic, centripetal transportation from Mrs. Bemis's stuffy house to the rambling backwoods hills, "pastures of softer grasses, . . . hummocks and hillocks, sedge and marsh," wherein, like a freshly re-centered Adam figure, he becomes reborn "at the edge of open country" (*We Fly,* 50). In October, Robert, with packed lunch, literally flees Mrs. Bemis's house on a Sunday, while the villagers dutifully congregate in church. Footloose in the fields, he finds his Sabbath becomes a sacrament of "witch-hazel with its curled yellow petals in full bloom," his choir the "windless air . . . filled with the faint half pleasant scent," his confessional an inner realization of self-fulfilling truth and abandonment into identity: "It was too late to change his mind. Too late to go back and change his clothes. Too late to change himself" (*We Fly,* 51). These passages show that, for Francis's main character, escape from civilization and self-preservation and discovery are synonymous.

While outdoors, Robert performs other symbolic acts. Robert's lunch by and nude immersion in a chilly stream constitutes a pristine Edenic sacrament, a neo-ecobaptism, and a re-visionary conversion for the reader from whose eyes the dark scales of humanity fall in the presence of the secluded scenery. This outdoor ritual occurs in a secluded spot surrounded by "birches and alders, with here and there a tall ash or elm or oak trailing the dark red of woodbine or the brighter reds and yellows of poison ivy." In describing this location, Francis emphasizes the "leisurely water," the "moat-like" stream that encloses a "small plot of green turf" and "the privacy of wood and water on one side and the equal privacy of open unfrequented country on the other (*We Fly*, 52). These Sunday afternoon escapades into the outdoors occur in quick succession—three times, as if feeding Robert's insatiable animal hunger—during which Robert's sense of human time dissipates. His presuppositions about humanity's centrality in the universe shifts, and the natural world approaches him in terms of grand art forms, music, and literature. Francis writes, "How long he had slept he had no way of knowing, and if he had known, how would the time have been measured?"; "Tree shadows were longer than they had been, and the slowly turning earth had moved him into one of them"; "It was like the return of a theme in music, or the rereading of a rich chapter" (*We Fly*, 54–55). In losing himself, by degrees, Robert discovers his place on earth.

If we turn from substance to structure, we discover grounds to cast *We Fly Away* not only as an unearthed bioregional narrative but as a pre-postmodernist piece of semi-autobiographical fiction.[3] The most noticeable and emblematic aspect of this novel's structure is that it contains no traditional textual breaks of any kind: no chapters, no sections, no headings. The entire one hundred fifty-six pages of text stream from cover to cover, from September to Memorial Day, as if to mimic the revolution of the earth and the blending of the unending seasonal cycles. The text's ceaseless structure reshapes, reorients, and redirects the reading experience so audiences cannot read a chapter, mark a page, put the book down, and conveniently pick up later where they left off. This structural feature alone invites readers to proceed against the grain of anthropocentric literary conventions. Francis's radically conceived structure requires readers to abandon themselves to its endless currents and elusive horizons so that the act of reading resembles the story's subject—a plunge into ancient rivers, a divergent hike through trackless wind, a time warp through three decades of seasonal weather patterns.

While this textual characteristic undoubtedly strikes some readers as a flaw on the part of a novice novelist, more likely Francis composed his novel in this form intentionally. As such, his resistant text invites us to read according to the cyclical flux of nature and urges us to surrender the compulsion to read according to human typological and publishing strictures. As a textual symbol, Francis's novel drapes a revolution of the seasons across the lives of four human characters and re-tunes readers to the placid maelstrom of climatic rhythms, toward temporal seamlessness and organic timelessness. Here again, Thoreau provides Francis with his philosophical and aesthetic credo. The twentieth-century nature novelist renders in textual terms the sylvan hermit's nineteenth-century rejection of calendars connected to "heathen deity," of time "minced into hours and fretted by the ticking of a clock" (Thoreau, par. 2). In a veiled allusion to Thoreau's nattering timepiece, Francis includes a frigid winter scene in which Robert, in an effort to thaw the "thin finger" of ice that has frozen a clock in Mrs. Bemis's kitchen, deposits it in the warm oven to thaw it out and, in so doing, literally cooks time (*We Fly*, 104–06). Examined broadly, the unbroken trickle and pulse of *We Fly Away*'s structural poetics strikes readers as evolutionary and revolutionary. The shape of Francis's narrative invites us to slip into what Leonard Lizak refers to as Francis's view of "cosmic time" to take a long look at the world (Lizak, 2).

The structural characteristics in *We Fly Away* are connected to its hallmark stylistics, which include its concentrated focus on bird imagery, bird references, and bird behavior—as the novel's title suggests.[4] While the novel's ubiquitous bird imagery may seem overdone to contemporary fiction readers, it nevertheless invites us to track Francis's narrative through the eyes of the creature as well as the eyes of culture. Years after *We Fly Away*'s publication, Francis is found still blending this avian brand of creature culture. "I live in a birdhouse," he tells Michael Hamburger, "and pick up my living bird fashion" (Letter to Hamburger). So heavy—or light—is the recurring bird motif throughout *We Fly Away* that it appears as if Francis, the sedulous constructor of the nest-narrative, daubs every textual crevice with its mud. Consider, for example, Mrs. Bemis's observation that Robert can "fly" around the house faster than she can. At one point, she confesses that her "flying" days are nearly over (*We Fly*, 14–15). Robert receives a vision of Mrs. Bemis as a "stuffed bird," and Henry insists that the "birds" of Bemis and Teal should "go to roost early" (*We Fly*, 26). Later, Henry describes Mrs. Teal as a "great blue heron" and is found "perching" in Robert's study. After Robert invites Henry into his bedroom and Henry refuses, Henry reacts like a tired raptor: "He flexed his arms, then spreading them like an eagle preparing for flight and bending back his head, he abandoned himself to a magnificent yawn" (*We Fly* 28–30).

At times, select verbs, phrases, simple similes, and scattered clichés amplify the avian: Robert observes Mrs. Teal as she "flit[s] about the kitchen." Mrs. Bemis, "with a chirp," reminds Robert to return a ladder to its place (*We Fly*, 34, 69). Robert notes how the "library brooded bird-wise over its books," and a town boy on a bicycle pedals past the Bemis home, his unbuttoned shirt flapping "like wings on either hand" (*We Fly*, 118). As Robert leaves Mrs. Bemis and prepares to live alone in his abandoned shack in the woods, he describes his future state: "I'll be free as a bird that builds its nest in an old apple tree" (*We Fly*, 155). As demonstrated here, a veritable flock of appropriate and well-chosen images, phrases, and references inhabits Francis's bioregional narrative. At the lexical level, his narrative engenders a breed of balance and reciprocity between the natural communities it describes and the human communities reading about Robert's experiences. While such heavy tropological insistence on bird imagery might strike twenty-first century audiences as wearisome, this language fashions the story into a bioregional narrative that mingles human and non-human biospheres.

The constant narrative execution and implementation of the ornithological functions both atmospherically and thematically. In the novel's opening sequence, the second paragraph finds Mrs. Bemis nidified at her "bird window," as bird-like in appearance but paradoxically far removed from being bird-like in actual behavior and nature:

> Here was a feeding tray, now deserted, but frequented in cool weather by sparrows and chickadees and sometimes a woodpecker. If they ever glanced up from their feeding, they may have sometimes seen an old, old woman on the other side of the glass, standing very still and peering down at them with little black eyes not unlike their own. So deep-set and intent were the eyes that *they seemed to focus on something nearer than the birds, and at the same time to be looking through them and far beyond.* She herself was scarcely larger than a large bird, with a bird's narrowness of shoulder. The face was convex, and the hair—a close-fitting transformation—was as smooth and neat as feathers about her head. But what was almost too bird-like was the hand resting upon the window-sash, withered and skeletal with age. (*We Fly*, 12, emphasis added)

The compelling aspect of this arresting textual invocation is the way Francis simultaneously associates Mrs. Bemis with and dissociates her from the wild birds that frequent her feeder. Clearly, her physical appearance resembles something "bird-like," though in excess of authentic "bird-ness." From the outset, Francis is careful to emphasize that Mrs. Bemis, as a domestic prisoner, can only slip closer to death behind a barrier of window glass, cut off from the natural world. Her chief characteristic remains her inability to see nature for what it is—like a bird. Instead, she is cursed, like a human sans perspective, to see only too near and too far.

In a similarly understated scene—the one from which Francis's novel takes its title—Mrs. Bemis floats unexpectedly into the kitchen and recites Psalm 90 to Robert. The impact of this impromptu speech mutes Robert within one of his signature prose codas of balanced atmosphere between the human and natural, the scriptural and environmental:

> "'The days of our years are threescore years and ten; and if by reason of strength they be fourscore years, yet is their strength labour and sorrow; for it is soon cut off, and we fly away.'"
>
> She stood looking into the distance. Then she turned toward Robert.

"'We fly away,'" she repeated in a higher voice. "Isn't that beautiful?" She made her lifted hand flutter, and her voice became still higher and softer. "'We fly—away.'"

The squares of sunlight on the kitchen floor trembled as the bare grape-vines outside the windows moved in the wind. The sunlight dimmed, dis-appeared, then flooded back. Two birds flew up from the feeding-board, and their shadows fluttered momentarily on the floor. (*We Fly*, 133)

As a narrative stylist, Francis excels at rendering his artistic style nearly transparent, as clear as the daylight by which his readers read, as weight-less and invisible as flight.

At a glance, 1948, the year *We Fly Away* was published, seems to have been a good year for blockbuster and forgettable novels.[5] That year, the Pulitzer Prize for Novels became the Pulitzer Prize for Fiction, and Zora Neale Hurston's *Seraph on the Sewanee* joined the fate of Francis's novel, though Hurston's work now enjoys greater critical attention than Fran-cis's does. In some ways, it might be said that the years following World War II were "meta-years" for the novel, producing a disparate array of fictional works that fall into diverse categories. Select scholarly studies in fiction for that time period, for example, Frederick J. Hoffman's *The Modern Novel in America: 1900–1950*, assign labels and qualitative cri-teria to these different types of emerging novels.[6] Interestingly enough, Hoffman's fourth and final category—"a growing concern with form"— describes *We Fly Away* perfectly. Within this "concern with form" cat-egory, Hoffman places the "novel of social purpose," a sub-genre that deals with the "burden inherited by the sensitive liberal who has out-lived the pertinence of his naïvely held principles" (Hoffman, 195, 201). Francis's story about a presumably gay nature poet's rejection of conser-vative middle-class values and his self-discovery in the sanctity of his natural environment matches Hoffman's specifications. As a hybrid novel whose compass touches literary trends and developments past, present, and future, *We Fly Away* tracks the quest of a one-man activist who willingly steps away from social expectation in order to carve out a finan-cially retrogressive but artistically, personally, and environmentally pro-gressive lifestyle. Its simplicity, its restraint, the quietness, the sensitivity, the stoic social pioneering—these characteristics combine to produce an inimitable bioregional narrative in whose pages readers find interwoven elements of the traditional *Künstlerroman*, eco-existentialism, and pre-postmodernism.

As a twentieth-century American *Künstlerroman*, *We Fly Away* blends art, humanity, and nature. Thirteen pages into the novel, Fran-cis stresses that Robert views himself as "a writer, though clearly not a prosperous one" and offsets this sense of the artist's identification with Mrs. Bemis's offhand scoffing about the way Robert "says he writes both poetry and prose," though nothing for her brand of magazine (*We Fly*, 13–14). Francis includes protracted passages about the way Robert sees the world in terms of art ("Byzantine sky"); the way he reads himself to sleep at night (Milton in *Palgrave's Golden Treasury*); how he writes at the library; ceaselessly revises his manuscript; mentally assembles a Fredric Jamesonian pastiche of Milton, Mother Goose, hymns, and Henry's drunken braying rendition of "How dry I am!"; and devoutly subscribes to the Wordsworthian practice of mixing writing and walking: "At such a time a poem might come to him—it seemed like a coming though it also seemed that his mind went out to meet it, and kept going out farther and farther to meet it. . . . Walking back to the house, he sometimes found himself stepping carefully as one carrying a full glass of water" (*We Fly*, 90–91). Woven in the novel's rotation of seasons is the artist's cycle of conception, composition, submission, and rejection. Francis depicts how Robert, within earshot of Mrs. Bemis's constant needling ("[H]e is supposed to be a writer"), surges with hope for publication but experiences the reality of rejection. This important plot element, however, doesn't mirror Francis's lived reality. After failing to publish two early collections of devotional and light verse, Francis's poetry manuscript *Stand with Me Here* was accepted in the spring of 1936, a thunderstroke of good news that threw him into such bouts of paralytic silence that he couldn't tell anyone for fear his luck might "evaporate or fly away" (*We Fly*, 120–21, 134; *Trouble with Francis*, 208). In this sense, Francis's novel tracks the progress and development of the nature of the artist and the art of nature.

The story of the struggling artist fuels the resolve of the eco-existentialist. During a frigid winter morning, Robert awakes to his situation as he eats breakfast in the house with his gloves on. "The absurdity of it more than the inconvenience suddenly filled him with rebellion," Francis writes. He depicts the suddenly introspective and restless Robert as "a sort of half-man, a semi-failure, in an old house with two old women equally fussy and narrow." Further, Francis reveals Robert's wish, if poverty is a must, to reside in a warmer climate: "If he had to be poor and cold, why couldn't he be poor and cold in a house by himself?" In Robert's idealistic mind, a place where art and nature coincide, success as a writer would mean freedom in the real world: "When his book of poems was published, he could afford to move to a warmer region. But whatever happened, he wouldn't stay here another winter" (*We Fly*, 105–06). The long view of Robert's trajectory, from perch to takeoff, reveals a smoldering pattern of thoughts and behaviors that tends toward self-determination, individualistic activism, a desire to shed social convention and constraint. Consider the moment Robert admits his shortcomings to Mrs. Bemis, who disapproves of a "man thirty-five years of age, in good health and of good education, who did not have a recognized place in the community, a decent income, a wife (and children if possible), and a home of his own" (*We Fly*, 13). Here, readers sense the degree to which Robert represents a torn and divided figure. He sees, in vision, Mrs. Bemis's dilapidated house as "an old ark of a ship" about to sail on a one-way voyage; he resents "routine," but he also initially aligns himself "on the side of convention and against the idol smashers." Soon, however, as a type of inmate to Mrs. Bemis, Robert exhibits the itch to "go and come without being noticed," which gives him a "feeling of privacy and independence," an urge that blooms and erupts at the novel's culminating "freedom" sequence (*We Fly*, 17, 20, 26–27).

At novel's end, which Winifred King Rugg misclassifies as "rather inadequate," Robert's transcendent drive fuses and transforms the narrative's four major themes (Rugg). This quartet of themes corresponds to the four seasons and four main characters. In May, directly following Robert's rejection from his potential publisher, he strides into the countryside, his "only thought" to "keep moving," his senses noting the "blossoms of the apple, the flowering quince," the "late forsythia" and the plowed furrows of farmland that curve "with the contours of the land" (*We Fly*, 135). "Freedom—that was the word," Robert thinks, as he hikes deeper into the uninhabited woods, fields, and streams. Without clear destination but with a lust for trackless wilderness motion, he considers how his recently rejected poetry manuscript had really been "an excuse" for "laziness and procrastination" and how he would now live "[i]n himself" (*We Fly*, 136). At the novel's crescendo, during which Francis begins paragraphs with the capitalized word "Freedom" three times, Robert discovers an abandoned "low-browed, shabby white house that seemed to have crawled up out of the bushes near the stream." At first, the shack "repel[s]" him but then makes him "feel at home" to the point that he tracks down the owner and investigates the possibility of renting it (*We Fly* 138–39). By literally flinging himself into the outdoors, Francis's protagonist takes his first strides toward liberation from domestic stultification and social imprisonment. In this sense, Robert's end becomes his beginning.

At this moment of consummation and conclusion, Francis's clear prose style barely hides a moment of bioregionalist experimentation and synthesis. On a subsequent ramble to the abandoned brookside shack, following the news of Henry's death, Robert undergoes a second ritualized nude baptism-by-stream in a memorable passage that in one narrative stroke recalls and conflates Henry's heron-like description of Mrs. Teal (the domestic), Henry (the homoerotic), Robert's poem "Clouds" (the eco-aesthetic), and Robert's final absorption into the wild world (the bioregional). At the stream's grassy bank, a "flapping of wings farther downstream . . . made him look to see a great blue heron lift itself into the air with slow wing-sweeps, and with its long legs trailing, sail away over the trees." This laborious but vivid depiction recalls the flighty Mrs. Teal. "Beyond the trees," the narrator follows, "beyond the bird, a daylight moon hung pale as a cloud—'farther than bird-flight, nearer than nearest star.'" Here, Francis inserts a line from one of Robert's poems, "Clouds," to fuse the artistic and the fleeting presence of the domestic. Immediately following these two allusive images, Francis abruptly inserts remembered or imagined dialogue spoken by Henry, the effect being that the reader sees the homoerotic, artistic, and domestic intermingled because of their proximity in the text. "Let's wait till the moon goes down," Henry says, presumably to Robert, though he never mentions why they should wait, nor what they should wait to do. An unidentified voice, presumably Robert in the past, asks Henry if he will be "sober" by then, which draws a "reproof" from Henry's voice: "Sober? Sober?" Then, in order to consummate the wedding between the homoerotic, domestic, and artistic in the natural world, Francis has Robert strip and lie nude in a nearby stream "like a martyr, on the stones" to "let the coldness flow over him." This fusion of the novel's four-fold themes provides Robert with the liberation from convention that he seeks: " 'Don't anybody tell me,' he panted as he staggered up, 'that I didn't do that in pure unadulterated freedom'" (*We Fly*, 147).

Shortly after this episode, Robert delivers his parting manifesto to Mrs. Bemis outside her home. Symbolically, he unfolds and flies her moth-eaten American flag in anticipation of Independence Day. His act of mock devotion and ascension repudiates society's expectation that he possess a "well-paying job and a wife and family and a place in the community." Instead, he chooses the secluded, poor man's luxury of being "free to be quiet and alone." On the day he leaves Mrs. Bemis, Robert passes the "grapevine . . . coming into bloom with its little greenish, inconspicuous flowers." Francis invokes this beginning-and-ending image to bring his narrative full circle. In doing so, he allows his blossoming protagonist and alter-ego to exit a cycle of human entrapment and enter a world of self-directed poverty, artistic transcendence, gender identity, natural reciprocity, sustainability, and ecocentric rapture (*We Fly*, 155–57).

If we consult only sales receipts, critical praise, and the re-issue of multiple editions, we cannot classify *We Fly Away* as the "success" that Richard Gillman labels it (Gillman, Intro. to *Travelling in Amherst*, xv). But perhaps the correct response to Francis's novel is not that it represents ordinary fiction but that it is extraordinary how accurately it reflects the facts surrounding his remarkable life. In an early journal entry, Francis stakes out his intention to own and inhabit a "little farmhouse off in the backwoods somewhere," an abandoned place he could "help repair" and in which he could "live and be happy" (*Travelling in Amherst*, 19–20). His autobiography records that this conceptually quaint farmhouse became a reality: a "ruined shack" off "Belchertown Road near Pansy Park," which Francis rehabilitated and inhabited before moving to Fort Juniper (*Trouble with Francis*, 196). It was this "old house by the brook" that an intrigued Robert Frost visited in the fall of 1937, where Frost "advanced the theory" that the abandoned domicile had been a rooming house for mill workers (*Frost: A Time*, 46). In ways direct and indirect, the story of Francis's life gives life to his story.

Additional details fuse Francis's biographical and bioregional narratives. His journals immortalize numerous descriptions of woodland retreats and brookside lunches that closely mirror the fictionalized excursions Robert takes in October: "In this open yet secluded place on the grassy bank I have eaten three Sunday dinners from a paper bag. . . . [B]luebirds . . . were flying from tree to tree. . . . The witch-hazel blossoms here a golden mist in the sunlight." In November 1945, three years prior to *We Fly Away*'s publication, Francis discloses his private thoughts concerning his performance as the author of a bioregional narrative: "This book . . . is based solidly on my own experience and observation: . . . [I]t bounded me helpfully, but it also restricted and bound me. It took me nine years to find the freedom of my material" (*Travelling in Amherst*, 45, 62). Elsewhere, Francis reveals that his ninety-year old landlady, Mrs. Boynton on Amherst's "North Prospect Street," served as the real-life model for Mrs. Bemis, and one seventy-year old Mrs. Kellogg was cast as Mrs. Teal in his "so-called fiction" that was "mostly not fiction at all" (*Trouble with Francis*, 208). To the bioregional narrative's characteristics we could add another: that it ceases to differentiate between fact and fiction, between bioregion and biography.

We Fly Away's ultimate distinction may be that it upends traditional notions of literary success and failure. In his interview with Elinor Phillips-Cubbage, Francis criticizes his "mistaken concept of writing novels" as a notion that literary works "had to be cooked up anew." In reference to *We Fly Away*, he adds, "This little piece of fiction (which I don't call a novel) shows how I overcompensated for my earlier mistaken notion of fiction" (Phillips-Cubbage 164, 170). In Francis's own words, we envision the putative failure he wrongly supposed his slim but sublime narrative was to become. He remembers becoming bogged down with "too much concern with technique" and with "point of view and stream of consciousness and Virginia Woolf" to the point that he couldn't focus on his story. "Years later when I wrote *We Fly Away*," he recollects, "I had come round to the other extreme and my story was all remembered experience and observation" (*Trouble with Francis*, 195). In a similar vein, he jots a nugget of writerly observation that depicts him as struggling to define his identity as a literary artist. In it, he gropes for the anchor of a genre he can call his own: "The poet is a spider, forever spinning. The novelist is a caterpillar, eating, eating great slices of life. But the poet spins his poetry out of himself, out of next to nothing" (*Travelling in Amherst*, 25). Despite such ambivalence, as a timeless contribution to the expanding library of bioregional narratives, *We Fly Away* has outdone, outgrown, and outlasted even its author's expectations.

In the 1976 Tetreault and Sewalk-Karcher interview, the subject of *We Fly Away* resurfaces. Here, Francis discloses why he did not write another novel: "It is my passion for economy with words and for using them to make something as well as to say something that explains why I have done so little in fiction. For fiction is inevitably verbose and like the 'confessional' poem it is mainly a telling of something that has happened rather than a making of something new" (*Pot Shots* 138). It thus becomes the overarching characteristic of the bioregional narrative to fly beyond the understanding of its author from the moment the author casts a backward glance and flies away.

EVOLUTION

Writing about Francis's prose is like trying to sketch a partially buried fossil. At present, over one hundred fifty environmentally themed "Home Forum" columns he wrote for the *Christian Science Monitor* between 1938 and 1954 remain uncollected in book form. Also, Philadelphia-based *Forum* (a *Current History* subsidiary that ceased publication in 1950), printed thirty-six of Francis's shorter "Country Comment" essays and eleven longer pieces. These uncollected literary artifacts gather undeserved dust in the archives of Amherst's Jones Library and on the campuses of Syracuse University, the University of Massachusetts-Amherst, and in private subterranean vaults and library basements. Add to this Francis's unpublished masterpiece *Traveling in Concord*, his uncollected journals, letters, satires, and criticism, and the scholarly world begins to sense how voluminous his output truly was—and how much of an injustice it is that this priceless prose goes unread and unappreciated.

"It is a delight to see Robert Francis again," the *Christian Science Monitor* Home Forum page writes in 1985, the year Francis received the Academy of American Poets' fellowship for "distinguished poetic achievement." The anonymous *Monitor* author remembers Francis's prose as having contributed to its pages "so richly in earlier days" ("The Poet"). Though the *Monitor* reminisced on the aesthetic wonder of his essays, Francis recalled the meager earth-to-mouth existence that forged the clarity, detail, humor, and sunlit sparseness of his most reflective and engaging pieces. "How poor

I was," he writes, "can be gauged from the fact that this 300 dollars a year was for years not only my most dependable item of income but also my largest" (*Trouble with Francis*, 22). Even taking into account the impact of inflation between 1938 and 2008, it is difficult to see three hundred dollars a year as an income substantial enough to sustain life. In many ways, writing provided food and survival for Francis, both body and soul.

Francis's uncollected essays present a contemplative view of the author's synergistic interaction with his natural surroundings. In a steady meditative tone, he muses on the manifold ways that nature molds his inner and outer worlds: whimsically, satirically, at times darkly. In "The Moon as Entertainment," for example, Francis details the protocol for hosting "a moon party," complete with guest chairs, binoculars, and the musical accompaniment of the "brook in the ravine" and "instrumental insects of a summer night." As a commentary on the 1950s age of television, "Television with No Antenna" touts the diversionary benefits of a star-gazing deck in exchange for the "little, faint-blue, flickering screen" Francis sees in all the homes he passes on his nightly walks. "For my television screen is the sky itself," he writes, "and the actors are stars, planets, meteors, moon, and occasionally the northern lights. . . . Compared with the spectacle far, far above their roof and its antenna, the small screen does seem small entertainment."

As amateur climatologist and mountaineer, in "Season Between" he meditates on the "season between seasons with familiar names," and in "Mountain Day" he revisits the college tradition of ringing a bell to release students into the mountainous October foliage when the weather, leaves, and hills erupt in a polychromatic "happy conspiracy." Vacillating between idealist and tragedian, in "Outdoors Indoors" he explores the "ideal state" of living "outdoors year round" and in "A Walk between Two Days" laments, "How many other November nights, as warm and misty and glowing as this one, had come and gone in other years while I had been indoors and sleeping?" One *Forum* "Country Comment" piece entitled "Nine Great Ones" treats one of Francis's favorite subjects, the aromas of the earth. Briefly, Francis lists the nine: (1) "the tonic, universal" smell of the sea"; (2) the forest's "cool moist exhalation of leaf mold"; (3) "evergreens hot in the sun"; (4) "the pungence of dry dusty land under the first down-rush of raindrops"; (5) "hay with plenty of clover and sweetgrass"; (6) autumn's fallen leaves and their "slow incense with the sun for fire"; (7) "clean cold air in winter"; (8) "fresh-ploughed land in spring"; (9) and "a bouquet of all widely-fragrant flowers." Such an epic mini-catalogue of scents functions as a kind of shield. "Hold them in your mind against a time of need," he declares. Anyone who reads these obscure but potent essays feels the curious sensation that the time of need of which Francis speaks was his past but is also our present.

A counterbalancing study of Francis's prose seems merited since those who knew him remember him as primarily a poet. Examined in conjunction with his poetic output, his diverse body of impressive but hard-to-access prose remains vital to defining his particular place in American letters. Throughout his scattered ruminations and reminiscences, Francis records the conflicted confessions of a man torn between identities. Was he a poet or prose writer? And why did he feel compelled to choose? In a journal entry (August 3, 1952), he appears to veer sharply toward poetry, wishing to "devote myself to poetry as unstintedly as mystic to God or as lover to beloved" (*Travelling in Amherst*, 74). However, five years later in Rome at an introductory social, during his year-long fellowship at the American Academy of Arts and Letters, he placed a card near his photograph and biographical materials that claimed he did not prefer poetry over prose, as other authors might. Finally, in 1980, inventiveness and compromise triumph in his mini-essay "Wordman" from *The Satirical Rogue Again*. In this essay he declares, "So let me be called a wordman and let what I write be called word arrangements. Though this or that critic might deny that I am a poet, . . . he could scarcely deny that I worked with words. As wordman I trust I would not threaten or irritate anyone" (*Pot Shots*, 166). With breezy satire, Francis plants a serious comment on writerly evolution, a critique of the needless distinctions and categories that scholarly publishing conventions impose on writers. This playful insistence that the writer in tune with his indoor and outdoor surroundings should eschew narrow existential categories and revel in hybridizations stems from Francis's observations about the natural world's tendencies toward evolution, interrelation, and indeterminacy of species. In "Market Hill Road," he compares a census of local poets to a census of trees: "When is a tree a tree? When is a poet a poet? Is a one-year seedling pine a tree or only the promise of a tree? . . . A sapling poet, such as Amherst sometimes produces in its institutions of learning, can be seen and counted easily enough; but what of all the tender seedling poets scattered among the grass?" ("Market Hill Road," 8).

Whether prose writer, poet, or wordman—or all three and more—Francis wrote and was changed. Daily, the word and the world transformed him into a new creature. His devotion to living simply and simply writing synchronized his activities with the rhythms of his environment. Those who read Francis, like those who read Thoreau, undergo similar yearnings for evolutionary change. In the reader's consciousness, a nagging dissatisfaction with the wastefulness of the contemporary world summons a desire to live more conscientiously, to consume less, and to generate more positive energy.

A rare document that bears record of Francis's life-long personal and artistic evolution is preserved in Leonard Lizak's "Robert Francis: A Trinity of Values—Nature, Leisure, and Solitude." This homespun but informative graph employs a purely functional title: "Biographical Chart by Robert Francis." The chart's curious arrangement represents the relationships between Francis's age, places of residence, style of living ("conventional" versus "unconventional"), occupations, and book publications. In the occupations bar, sundered by a small block labeled "army," Francis comments on his status as an evolutionary creature. In assembling the chart, he appears to track his growth from stage to stage, as if he slowly progressed from egg, larva, to pupa, and finally mature adult. Though Francis's penmanship in his biographical chart remains only partially readable, he clearly classifies himself not as a writer only, but as a "writer & something else." The "something else" appears to be "music t.," or music teacher, given the many years Francis earned extra income as a local violin teacher. Francis's master's thesis, "Expression in Violin Playing," completed his degree requirements at Harvard, and his violin, along with surviving music and compositions, remains on display at the Jones Library in Amherst. Despite this formative musical background, his autobiography captures the inner stirrings that propelled him toward further evolutionary stages of growth, from strings to wings: "After the dislocations of war, I decided to be a full-time writer. . . . Violin teaching had been a stop-gap. . . . I could go hungry in a good cause" (*Trouble with Francis*, 77). Francis's biographical graph and autobiographical recollections chart his final blossoming from meta-writer to full-time writer. An arch of ceaseless struggle toward fullness and fruition, like that experienced by praying mantises he so often observed, kept him grounded and gave him flight.

Ecocritics early and late have isolated evolution as a theoretical concept that overlays literature and the earth.[7] As contentedly as Whitman's grass and redwood tree, or any lungfish or peppered moth of Birmingham, Robert Francis evolved as a means of survival. An early journal entry (August 9, 1932) preserves his philosophical credo regarding humankind's need for change, adaptation, and continuation. "All fixity is a dream," he offers, "and when man wakes from his dream he wakes into disillusion. Let us base our lives on the ceaseless flow and change of things. . . . Let us not be disturbed at finding ourselves changing or our friends changing; let us rather be disturbed if we do not change" (*Travelling in Amherst*, 18). As an evolutionary species, Francis remained constant only in that he never ceased to change.

In 1979, the city of Amherst established October 21 as "Robert Francis Day." In his honor, the city dedicated a footbridge that still spans Mill River. From a more global vantage point, contemporary readers and walkers might see Francis's literary legacy as a rude but sturdy natural arch from *Walden* in nineteenth-century America to the sea of turmoil in which information-laden citizens of the world flounder now in the twenty-first century. In a time of hurried comings and goings, of harried indecision, he invites those seeking respite to pause over the gorge between destinations and definitions and, suspended in mid-air, to be held in what the senses behold.

Photo of Robert Francis from *Longman Anthology of Contemporary American Poetry: 1950-1980*.

Run Poetry

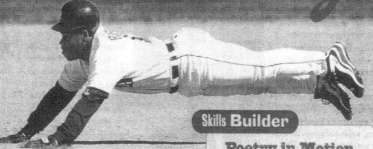

"Delicate, delicate, delicate — now!"
Speedster Bip Roberts swipes another base.

The Base Stealer

Poised between going on and back, pulled

Both ways taut like a tightrope walker,

Fingertips pointing the opposites,

Now bounding tiptoe like a dropped ball

Or a kid skipping rope, come on, come on,

Running a scattering of steps sidewise,

How he teeters, skitters, tingles, teases,

Taunts them, hovers like an ecstatic bird,

He's only flirting, crowd him, crowd him,

Delicate, delicate, delicate — now!

—Robert Francis

Skills **Builder**

Poetry in Motion

Now it's your turn to write a sports poem. Follow the instructions below.

♦ Choose the type of athlete you want your poem to be about. In the poems at left, the poets chose "Base Stealer" and "Outfielder." You can choose a player or position in any sport you like — pitcher, quarterback, goalie. It's up to you.

♦ Write the player you've chosen across the top of a sheet of paper, and down the lefthand side of the same sheet. This will be the title of your poem, *and* will provide the first letter of each line.
Look at the example below.

Catcher

Can snatch a speeding ball from thin air.
Absolutely able to throw stealers out.
Tells the pitcher: "High, low, or away."
Convinces the ump they're all strikes.
Hits, too!
Errors? Never!
Runs to first as fast as lightning.

♦ Following the example, write your own poem. Use as many descriptive phrases as possible. Try to capture the essence of the athlete you're writing about.

Scholastic Scope Feature on "The Base Stealer."

Robert Francis, a Poet Hailed by Frost, Dies

AMHERST, Mass., July 15 (AP) — Robert Churchill Francis, once described by Robert Frost as "the best neglected poet," died Monday in Cooley Dickenson Hospital in Northampton. He was 85 years old.

Mr. Francis had lived simply in a two-room home here for more than 40 years. In 1984 he received the Academy of American Poets Fellowship Award for distinguished achievement.

Born in Upland, Pa., Mr. Francis moved to Amherst in 1926 shortly after graduating from Harvard University. He taught high school for one year, then devoted his life to writing poetry.

"My speciality has been not to earn much but to spend little," Mr. Francis told The Daily Hampshire Gazette in a 1981 interview.

The publication in 1936 of his first collection of poems, "Stand With Me Here," brought him an invitation to be a fellow at the Bread Loaf Writers Conference and two years later he received the Shelly Memorial Prize.

He also served as Phi Beta Kappa poet at Tufts and Harvard universities and in 1957 received the Rome Prize Fellowship given by the American Acadmy of Art and Letters.

The last of his dozen books, "Travelling in America," a collection of his journals, was published last year.

The University of Massachusetts Press Juniper Prize for Poetry was established in Mr. Francis' honor and named after his small, tree-shrouded home that he called "Fort Juniper."

New York Times Obituary

This Certifies That

Robert Churchill Francis

was

Baptised

into the fellowship of the

West Medford Baptist Church

At West Medford

On the 12th day of April

In the year of our Lord, 1914

Elew F. Francis

Pastor

Watchword:

I am the good shepherd, and know my sheep, and am known of mine.

John 10:14

Do the hard thing.

Francis's Baptismal Certificate.

By gradation hills are worn down and valleys filled
up until the base level is established. Sometimes
this means a peneplain containing a single monadnock
or solitary mountain of more resistant material than
its surroundings and occuring farthest away from the
drainage streams.

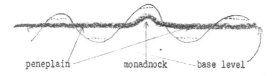

peneplain   monadnock   base level

II. Weathering (cf. gradation)

By weathering the corners of a boulder break off first   *Talus is*
(called spalls), and if on a steep slope slide down and   *the pile*
accumulate as talus(large pieces) or scree(small pieces).   *Slide rock*
Landslides are also of this nature.   *are the*
                                                            pieces.
Agents of weathering vary in effectiveness at different
places and at different times in the same places. But
some agent is always active. Two factors determine speed
of weathering: (1) composition and structure of rocks,
(2) conditions to which rocks are exposed. Illustrations:
(1) sandstone vs. granite, (2) adobe in dry vs wet climate.
Two groups of agents: (1) mechanical, (2) chemical.

Mechanical agents.
1. Frost. Works by expansion. Water expands 1/10 its
   volume, and exerts pressure of 150 tons a square foot.

Lecture 4   Oct. 9

1. Frost (continued) All rocks have some porousness
   for water and frost to work on. Cracks split open
   and rough edges break off. Some sandstones may add
   1/8 to weight by absorption of water. Sandstone
   obelisk in Central Park wore so it had to be painted,
   though it had endured for centuries in Egypt. Among
   tombstones, sandstone wears fastest, marble last longer,
   granite longest. Granite is granular rock made up of
   (1) quartz, (2) feldspar, (3) hornblende, and(4) mica.
   First two are light-colored; third is dark. Gruss(?)
   is broken-up granite, done largely by frost.

   Talus slopes also frost work.

Francis's Geography Class Lecture Notes (1944).

Geogra hical locations of poems many of which are in
Stand With Me Here

Hay Heavinesss	South Amherst
Hay	" "
Onion Fields	Connecticut Valley
Rana	Mt. Toby
Bronze	Hop Brook, South Amherst
The Outgrown Garden	Amherst
The Runners	College Hill, Amherst College
February Snow	Johnson Chapel, " "
Apple Gatherers	Atkins Orchard, South Amherst
Artist	South Amherst
Prophet	" "
The Celt	Amherst College campus
Hell	South Amherst
First Sister	" "
Firewarden on Kearsarge	Mt. Kearsarge, New Hampshire
Legend of Orient Point	Long Island, New York
Hermit	Greece
Shelley	Italy
While I Slept	West Medford, Mass.
Night Train	South Amherst
Fall	" "
Valhalla	Vermont
Blueberries	Chorcorua or Kearsarge, N.H.
Breaking the Apple	Cushman
The Plodder	Pelham
The Stile	Hadley Road, Amherst
Dwight	South Amherst
The Gardener	" "
Two Women	Breadloaf, Vermont
Solitaire	Cushman
Dark Fire	Fort Juniper
The Sound I Listened For	Cushman
Altitude	Hadley Road, Amherst
The Hay Is Cut	Cushman
The Good Life	Connecticut River Dam, Holyoke, Mass.

Francis's List of Poems and Related Geographical Locations.

Biographical Chart from Leonard Lizak's M.Ed. Thesis (1966).

· ·

A TRIBUTE TO W. H. AUDEN
AND HIS VOCABULARY

Whether it's deasil, cancrizans, or widdershins,*
Auden always finishes what he begins.

Oh, you can count on him to carry out his plans
Whether they're deasil, widdershins, or cancrizans.

All his "i"s he dots, he crosses all his "t"s—he'll
Fit them in: cancrizans, widdershins, or deasil.

> Friend of felicity, archfoe of flaw,
> Whose word (deasil or widdershins) is law,
> Of Auden (Wystan Hugh) we stand in awe.
> —ROBERT FRANCIS

*Deasil: from east to west.
 Cancrizans: crabwise or backward.
 Widdershins: in a contrary direction.

W. H. Auden Satire from *The New Yorker* (1956).

A black iron fence closes the graves in, its ovals delicate as wine stems. They resemble those chapel windows on the main Aran island, made narrow in the fourth century so that not too much rain would drive in . . . It is April, clear and dry. Curls of grass rise around the nearby gravestones.

The Dickinson house is not far off. She arrived here one day, at fifty-six, Robert says, carried over the lots between by six Irish laboring men, when her brother refused to trust her body to a carriage. The coffin was darkened with violets and pine boughs, as she covered the immense distance between the solid Dickinson house and this plot. . .

The distance is immense, the distances through which Satan and his helpers rose and fell, oh vast areas, the distances between stars, between the first time love is felt in the sleeves of the dress, and the death of the person who was in that room . . . the distance between the feet and head as you lie down, the distance between the mother and father, through which we pass reluctantly.

My family addresses "an Eclipse every morning, which they call their 'Father.'" Each of us crosses that distance at night, arriving out of sleep on hands and knees, astonished we see a hump in the ground where we thought a chapel would be . . . it is a grassy knoll. And we clamber out of sleep, holding on to it with our hands . . .

Robert Bly's "Visiting Emily Dickinson's Grave with Robert Francis" from *Painted Bride Quarterly*.

Photos of Fort Juniper and "Old House by the Brook."

```
        1974 Income through Dec. 31.


    Earned Income                          Time of Visit

    $  50  Worcester Poetry Assn.          Nov. 1973
       150  Quinsigamun College            Nov. 1973
        75  Lincoln-Sudbury High           May  1974
       125  Dracut High                    May  1974
       200  Assumption College             May  1974
       250  NYSEC(Binghamton)              May  1974
       125  AIC                            Nov. 1974
        25  Macmillan
    212.72  Wesleyan Press
    901.23  UMass Press
    ----------
    2113.96  Total

    Savings Bank Interest
         779.44

    Social Security Benefit Checks

    #133.70
     112.80
     112.80
     121.50
     369.000
     126.00
     126.00
     126.00
     126.00
     126.00

     1479.80   Total for year



    Savings Bank Balance        $13,250.51

    First National Bank Balance     $891.37


    My town property tax has been paid, first half $16.48,

      second half 34.59.  Total $51.07   (This is the third year
                                           I have had a rebate of $350.

    US income tax  $238 plus "penalty" of $16.57. Total $254. 57

    Social Security Tax  $303

    Total of the two $557.57
```

Personal Income Record (1974).

Photos of Francis's "Stone God" at Fort Juniper.

Photos of Francis Memorial Bench and Bridge.

5
Ecospirituality and Ecopolitics

God slays himself with every leaf that flies.

—*E. A. Robinson*

Do they not see that we are bringing destruction upon the
land by curtailing it of its sides?

—*The Koran*

IT IS DIFFICULT, AND PERHAPS unnecessary and counterproductive, to try
to separate Francis's politics from his views on spirituality. Throughout
his lifetime, Francis returned to the subjects of religion and politics in
his writing, as if searching for footholds in the shifting existential sands
of modernism. While he peacefully gravitated from his father's North-
ern Baptist theology, his pacifist feud with the decimation wrought by
the twentieth century's cycle of warfare and environmental slaugh-
ter raged until the day he died. As a constantly emerging American
ecopoet, Francis coupled an "open" spirituality (a curious pastiche of
Christianity, nature worship, Buddhism, Hinduism, and other Eastern
world views) with a committed, more "closed" political stance (paci-
fism, non-violence, conscientious objection, activism, and dissidence).
For this reason, a more complete picture of Francis's ecopoetics must
include an examination of those poems that loosened his spiritual out-
look and those that hardened his political resolve.

THE SEEDS OF FAITH

In the winter of 1920, Francis, then an eighteen-year-old Harvard
undergraduate, returned home due to a "persistent pain" in his "lower
right side" (*Trouble with Francis*, 178). That Easter, after an extended
convalescence, Francis's father, Ebenezer, expressed his wish that
Francis should dedicate his life to God by performing pastorly
duties—a "thank offering" for his recovery, as Francis remembers it in
The Trouble with God (112). According to his recollection, Francis's

110

first and only experience as a deputized country pastor involved a ride on a "train to Westfield" where a "farmer in a jalopy" drove him to a "small pastorless church" in which Francis preached a sermon entitled "Growth" (*Trouble with Francis,* 179). Throughout these *ad hoc* ecclesiastical performances, Francis found himself "challenged" by the requirement that he say grace before meals with those who hosted him, and on his return trip, his hostess (along with a young boy and the boy's mother) proposed an impromptu excursion up the propinquant Sinai of Mount Tom, during which Francis "gallantly" paid all four travel fares. Consequently, at home in West Medford he found that his transportation expenses had exhausted the ten dollar honorarium the church had allotted him. Thus, he completed his spiritual tour of duty, no more financially well off than when he started out. "This was the nearest I have ever come to the ministry," he recorded, "even as my appendicitis is the nearest I have come to death" (*Trouble with Francis,* 179).

Though his father's faith, like Harvard, never captured Francis, the title of his singular sermon—"Growth"—could not have proven more prophetic in describing his evolutionary drift toward embracing a John Clare-like "religion of the fields." While Francis was certainly a man of faith, precisely what his faith consisted of—and how he wove its manifold hues into his writing—remains a complicated subject to address. Most useful, perhaps, in exploring the interface between nature and theology in Francis's poetry would be a topically divided reading in conjunction with John Gatta's notion of "ecospirituality" or "nature spirituality" (Gatta, 7). Ecospirituality, Gatta writes, "frequently presents itself in American literary culture as a supplement rather than a surrogate for revealed religion" (Gatta, 5). Within his broadly defined framework, Gatta sees "Nature" as a force that "points toward a protean, elusive, highly subjective phenomenon that is inextricably wedded to cultural assumptions and human imaginings" (Gatta, 9). In surveying Francis's spiritually centered ecopoetry, twenty-first century readers discover a clearly protean but somewhat elusive belief system, an evolving range of religious viewpoints that can be grouped into three categories: traditional, alternative, and skeptical.

By ecopoetry of a more "traditional" spiritual orientation, I mean that Francis employs historically familiar Judeo-Christian figures, tropes, and metaphors to explore the connection among spirituality, the natural world, and rural environments. Within this group of ecopoems in the "traditional" category, however, many attitudes and intents surface, ranging from reverential to satirical, from awe to apathy. In "The Curse," the speaker declares, "Hell is a red barn on a hill" out of which "on a Sunday morning" a voice can be heard "God-damning cows / At milking" (1, 6, 8). With blunt irony, the satirical speaker inverts the supposedly placid rural scene and paints it with the realism of neo-scriptural invective. In "The Reading of the Psalm," however, a palpable votive atmosphere pervades the speaker's description of a grandmother who meditates over a thunderstorm in the attitude of someone at solemn mass. The grandmother almost prayerfully watches how the "lightning parts the sky" as though she were observing the "lighting of a candle" then turns "from the storm as one might read a psalm / And look away and slowly close the book" (14–15, 27–28). "Psalm" adopts archaic verb forms and reads like a formal sermon, with nature as oracle: "And he shall be / Like a tree / . . . That bringeth forth / His fruit in his sea-son" (5–6, 12). In "Able," Francis puns on the name of the Biblical son of Adam in order to praise Abel's devout agrarianism and connection to the earth: "Both to converse and to conserve / Frugal and fruitful, passionate-patient / Never mistaking facility for felicity / Salesmanship for craftsmanship / Skeptical of commendation of condemnation / This is the man" (1–6). In many ways, Francis appears to uphold as many Biblical traditions as he overturns, delivering salutations of fiery praise as well as the hellfire of satire.

At times, Francis's satire proceeds with subtlety that borders on invis-ibility, while his rapturous ecstasy explodes with the force of a cock crow. In "Prophet," the speaker describes a pipe-smoking elderly male figure who with "appraising, practiced eye" steadily "scans the sky" and with subtle arrogance attempts to pronounce "what God hath wrought," his "foot on an old log," the smoke from his pipe rising up to "join the fog" (1–6). In the end, the speaker appears to critically assess the preten-tions of the anonymous and self-proclaimed country prophet and his putative power to alter the patterns of climate and atmosphere: "In all fair weather he smells rain / So doggedly I wonder whether / He does not inwardly complain / that foul days sometimes breed fair weather. / But under this inscrutable sky / What can a prophet prophesy?" (10–15). "Prophet" questions humankind's supposed superiority over nature and anticipates many foundational ecocritical critiques of cultural dualisms that favor human-nature divides.[1] "Prophet" and other "traditional" eco-poems, such as "Thistle Seed in the Wind," appropriate Biblical language as a means of re-writing anthropocentric scriptural injunctions: "Pioneer, paratrooper, missionary of the gospel seed, / Discoverer, skylarker, par-able of solitude, / Where is the mathematics, wisp, to tell your chance? / If when you fall you fail, / Are lost at last and die, / At least you will have made the great voyage out, / Your sun-slanting sail alone on the blue ocean-sky. / Hail, voyager, hail!" (1–8). Such ecstasy, coupled with such brooding dejection, suggests that Francis, through the act of writing poetry, understood what it meant to take the great voyage out into the world and into himself: to find, lose, and re-discover God in the post-Edenic wilderness of the earth.

In Francis's personal writings, readers grasp the authenticity of his struggle to negotiate traditional modes and forms of worship via more non-traditional ecocentric avenues. In his autobiography he recounts his first "religious experience" in Greenport when, as a young boy, he "rushed from the room in tears and ran home" after the singing of the climactic verses of "Come Thou Fount of Every Blessing," though he remembers being baptized as a young man "without any undue emotion" (*Trouble with Francis*, 139–40). In his journals, he emerges as a subversive iconoclast who, in his father's absence, attends the sermon of the substitute preacher, Mr. Pierpont, and drapes the pulpit in blazing garlands of unauthorized October leaves, blossoms, and flora. "The front of the church was swimming in golden light," he remembers, "maple leaves scarlet and yellow, gold-edged coleus, gold and burnt-orange calendulas, green ferns and rich-leaved begonia." The "joke," as he recalls it, stemmed from the contrast between the bloodless figures in his local congregation and the vibrant colors of his bioregion: the "hideous . . . stained glass windows" shedding "just the right light to bring out the mellowness of the colors" and "Mr. Pierpont's gaunt white and black wreathed" in an "autumnal riot." Ultimately, he remembers with satisfaction, "Those who had come to church out of duty forgot duty and enjoyed themselves. We all looked as if we were listening to the sermon while we drank in great draughts of color. Even the sleepers weren't quite so sleeping. And so a smile kept playing over my face—right through the sermon" (*Travelling in Amherst*, 6–7).

While Francis depicts himself as a soul renegade and eco-iconoclast, he does not spurn all visionary experiences, least of all his own. With reverential sobriety, he details how on a "rainy afternoon" he experienced "the nearest to a vision or revelation" he would ever receive—the visual and mental epiphany of a "little farmhouse off in the backwoods somewhere" that he could "help repair" and in which he could "be happy" (*Travelling in Amherst*, 19–20). In his autobiography, he describes the "ruined shack" (or "old house by the brook") that he "dreamed of rehabilitating or at least of occupying spiritually" and how that dream became reality (*Trouble* 196). Clearly, while he took hammer and tongs to the putative delusions of organized religion, he still dreamed dreams and saw visions.

In using the appellative "alternative" to designate a subset of Francis's spiritual-oriented ecopoems, I mean that he devotes a significant portion of his writing toward Eastern philosophies and religions, as well as perspectives that at times appear henotheistic, pantheistic, panentheistic, and/or neo-pagan.[2] In "Fortune," an "older man without apology" reclines "where water / Mirrors the green of grass" and portends the imminence of "travel" in the markings on his "neighbor's foot," which the older man holds "like a palmist in both hands" (2–5). In "The Outgrown Garden," Francis observes "Summer" in the act of "playing God" (7–8). In "Hay," he sees the "haymen" who "ride to the barn in waves / Of hay" as "New England Neptunes" (12–13). How seriously Francis regarded mysticism, paganism, and forms of nature worship remains unclear, though in his autobiography when he recalls the year that Ballantine Books included six of his poems in Rolfe Humphries's *New Poems by American Poets* (1953), he writes, "So important it was to me to be in that volume that during the period of editorial selection I just about turned to prayer and magic" (*Trouble with Francis*, 84). Elsewhere, he remembers his practice of sunbathing as a "religious rite and ritual." He explains, "If the sun lacked many of the attributes of the Christian God, his blessings depended on no act of faith to be effective" (*Trouble with Francis*, 56). Part mystic and sun-worshipper, Francis followed the Green Fates.

In another poem, "The Dandelion Gatherer," Francis turns a female gypsy figure into a rural priestess or enchantress "[b]ulging in petticoats in May . . . / Barefoot or in bursting old shoes, her hair / Bandanna'd and her ears hooped down with gold" (1–3). This vagabond prophetess of the high wilderness bears a "gunnysack with no one knows / How many thousand martyred golden heads—Phoebus Apollo, Dionysius, Christ—/ All lost, all plundered, severed, and all saved: The gleaming wineglass and the golden wine" (4–8). In these compact and elliptical lines, Francis's speaker fuses Christianity, polytheism, the vegetable and human biospheres, all doctrines of deliverance and damnation, and in the amen of his final line his lyric trickles into a gleaming chalice of homemade dandelion wine. Here, dandelion wine, which paradoxically combines the decadent and sacramental, symbolizes Francis's view of spirituality as something that gradually condenses, ferments, and distills over time, a process of fusion and synthesis. In Francis's case, home-brewed dandelion wine provided more than poetic metaphors. In a letter, from Raymond Adams to H. Leland Varley, Adams asks Varley to tell Francis that he remembers Francis's "dandelion and elderberry blossoms wines" and that "the years have been flavored pleasantly" (Adams). In a letter to Samuel French Morse, then at the Mount Holyoke English department, after declining Morse's invitation to speak on "The Changing Image of Robert Frost," Francis discusses his latest batch of "dandelion wine, alias fruit-and-flower cordial," and how it "still holds out" (Letter to Morse). That Francis considered the distillation of the world's religions from the standpoint of a self-taught nature observer surfaces in his journal (February 1935), where he notes, "One complex problem has haunted me more than any other these years that I have tried to write: the need and possibility of a new spiritual synthesis to take the place of the religions"

(*Travelling in Amherst*, 40). In the power and glory of poetry, Francis lifted nature's holy liquor to his lips and sought saintly intoxication in the fusion of world and spirit. Poetry was the embodiment of his wish to share this rapture with his readers. In the preface to his *Collected Poems*, he invites readers to "happen on an oracle" (*Collected Poems*, XX).

Mingled within this ecopoetry of varied "alternative" castes, a tattered golden thread that shows an affinity for Eastern philosophies, religions, and thought rambles through the pages of the Francis canon. In most if not all cases, these poems probe the tenuous connection between human and non-human natures and direct a meditative awe toward nature as the primary source of enlightenment, truth, and wisdom. "T'ang Poems" reads, "These words are cool as old tombstones / In a lost graveyard under pines; / Brief as chiseled epitaphs / Telling of a time long dead" (1–4). "Waxwings" begins, "Four Tao philosophers as cedar waxwings / chat on a February berrybush / in sun, and I am one" (1–3). The ambiguity in "one" communicates the way Francis includes himself as one of the bird-philosophers even as he achieves a placid inner sense of organic one-ness or wholeness in the act of beholding the birds. "Was this not always my true style?" he asks the world, concluding, "To sun, to feast, and to converse / and all together— for this I have abandoned / All my other lives" (10–12). In establishing the poet in a relationship of equality with, rather than superiority over, the waxwings, Francis drifts from an exploitive Euro-American *episteme* toward an Eastern view toward nature. Yi-Fu Tuan has contrasted the Chinese and Euro-American views toward the environment accordingly: "What is the fundamental difference between European and the Chinese attitude towards nature, that the European sees nature as subordinate to him whereas the Chinese sees himself as part of nature" (Tuan, "Discrepancies," 92). Also, by referring to multiple lives in his last stanza, Francis gestures toward re-incarnation and Hinduism.

"The Buddhist does not speak of God," Francis observes in *The Trouble with God*, "yet no religion probes more deeply into the human psyche than does Buddhism" (*Trouble with Gold*, 38). Buddhism and Buddhist imagery form the foundation of several of Francis's poems, such as "Peace," in which the speaker, while aware of "other men . . . crumpled to their knees / Broken and bloody underneath the whips," contemplates the meditative serenity in an ancient blue porcelain sculpture in a museum: "Bronze Buddha, Buddha of stone / Beaming and benignant as the moon / . . . Peace within peace, the peace of Buddha's smile" (1–2, 5, 13–14). His poem "Rana" breathes the spirit of Buddha into a frog. "Buddha never sat so still," the speaker chants, "Nor any graven Buddha sits / . . . So still as this frog on the stone / . . . Bronze-of-rock, / Water-green, sun-jewelled / He contemplates the All" (1–2, 4, 6–9). In associating inert amphibians with Buddhism, Francis appears to have generated a brand of nature poetics that predates what Laird Christensen calls the "pragmatic mysticism" in Mary Oliver's poetry, particularly a poem such as Oliver's "Toad" in which a toad that "might have been Buddha" does not "move, blink, or frown" or let a tear fall "from those gold-rimmed eyes as the refined anguish of language pass[es] over him" (Christensen, 135).

Though the amphibian in "Rana" is a frog, in an autobiographical section that catalogs the lives of his many "nonhuman associates," Francis mentions a toad he named "Todo" who "haunted the purlieus" of Fort Juniper, "as moveless as Buddha," perpetually "contemplating the All" (*Trouble with Francis*, 74). For weeks, Francis claims, he kept a daily diary that recorded Todo's movements: "The armistice was announced by radio at 9:00 this evening. At 10:00 the President spoke. A few minutes later I looked outdoors and there on the stone, . . . sat Todo, as if he too had been listening to the President" (*Trouble with Francis*, 74). Richard Wilbur, in his introduction to Francis's *Butter Hill*, calls Francis a "New England poet with Japanese virtues." Wilbur qualifies his observation: "We seldom have a sense that a garrulous someone called 'I' is standing between us and select phenomena. . . . Francis is enough *out there* to know that a nuthatch sees him as upside-down. . . . The poet seems at most a witness; when the images . . . appear on their own initiative . . . and some mood or thought gradually transpires" (Wilbur). In pockets of light throughout the Francis canon, East meets West—or, more accurately—Francis writes as if the division between East and West never existed.

Alongside his "alternative" compositions, Francis's "skeptical" ecopoems, like his more traditional spiritual pieces, adopt a multiplicity of stances, as if exploring the infinite degrees of doubt. Some ecopoems in this category fire devastating salvos at institutionalized religions, framed within all-encompassing ecological truths. "History," a pun-laden four-part satire on anthropocentric versions of history, pits the "Holy See" versus "the unholy sea / Scrubbing earth's unecclesiastical shores" (22–23). The stridently anti-Catholic "St. Brigid's" brick-by-brick dismantles the church of the same name, which Francis must have viewed from the window of his Cowles Lane apartment in Amherst during his brief residence there in 1975. In this poem, superior to the "little priests" who "scheme to dislodge or utterly destroy," the "Doves / who come and go to church / without ever going to church at all" from the belltower behold the "actual / undoctored unindoctrinated world, the non-theological landscape" (10–12, 28–30, 35–36). "Only as the spirit moves / with merely / an initial flap of wings," the doves bend "to their own dove devices / that massive immoveable institution," compelling the speaker to exclaim, "How beautiful / this dove indifference to Rome / . . . also / to mass, confession, Thursday / bingo and to the North Italian style" (41–42, 49–51). In the ecological imagination, Francis's Juvenalian satire marries the Magna Mater and the Error Pater.

In more than one poem, Francis sketches a thinly-veiled self-portrait of his middle-way approach to securing spiritual satisfaction in the ecosystemic phenomena of the earth as well as more cosmic points of reference. In these milder skeptical poems, he emerges as neither staunch believer nor heretic but both. In "Old Man's Confession of Faith," the speaker writes, "The blowing wind I let it blow, / I let it come, I let it go. / Always it has my full permission. / Such is my doctrinal position" (1–4). Unlike the wind in the Gospel of John that affirms the operation of the Holy Spirit, Francis's wind serves as bi-directional apotheosis toward the salvation of remaining undecided and uncommitted.[3] "Epitaph" labels the speaker "Believer and unbeliever both / For less than both would have been less than truth" (1–2). "Portrait" places the speaker near "something verging on humility" between "infinite learning" and "spiritual malaise" whose variations display "as many phases as the moon's" (1–2). "Having embraced the heresies one by one," the speaker concludes, "He can afford now to affirm the faith / . . . Like a cold peak he rises out of mists / which he once loved and, moonlit, still might love" (6–7, 11–12). Thus Francis's skepticism ranges from the acerbic to the amiable.

Two final but crucial ecopoems in the skeptical category, "The Orb Weaver" and "Nothing Is Far," illustrate the polar extremes to which Francis found himself driven at times, between a black brand of ecological antinomianism and a sublime agnosticism. "The Orb Weaver," which has been compared to Frost's "Design,"[4] describes a "serenely sullen" orb-weaver spider (family *Araneidae*)—"devised of jet, embossed with sulphur / Hanging among the fruits of summer"—that waits with sinister patience to skewer its prey with a "sudden dart" and "needled poison" (2–4, 7, 10). Structurally, four tercets are divided by this single maverick line: "And in its winding-sheet the grasshopper" (7). In his unpublished treatise on poetic craft, *Francis on Poetry*, Francis reveals that this structural curiosity serves an intentional purpose, in that the single line physically represents the grasshopper mired in a spiderweb, awaiting its execution (*Francis on Poetry*, 75). In the final tercet (which completes the unlucky total of thirteen lines), the speaker shakes his fist at the sky: "I have no quarrel with the spider / But with the mind or mood that made her / To thrive in nature and in man's nature" (11–13). Andrew Stambuk calls this conclusion "unequivocal in its view that a malevolent creator . . . from whom darkness and destruction emanate . . . is accountable for the predation that the poem describes" (Stambuk, 550). While Stambuk's assessment certainly rings true in one sense, readers no doubt register the irony in Francis's bellicose standoff with the God of Nature: to quarrel with the supernal mind that fashioned arachnid assassins is to acknowledge its power and presence, however inscrutable. In negating God, Francis affirms God's existence.

In "Nothing Is Far" (originally titled "No Pantheist"), Francis assumes a more mellow relationship with nature's baffling peculiarities. This poem's streamlined iambic tetrameter lulls the reader into a state of dreamy semihypnosis:

> Though I have never caught the word
> Of God from any calling bird,
> I hear all that the ancients heard.
>
> Though I have seen no deity
> Enter or leave a twilit tree,
> I see all that the seers see.
>
> A common stone can still reveal
> Something not stone, not seen, yet real.
> What may a common stone conceal?
>
> Nothing is far that once was near.
> Nothing is hid that once was clear.
> Nothing was God that is not here.
>
> Here is the bird, the tree, the stone.
> Here in the sun I sit alone
> Between the known and the unknown.

As timeless and compelling as this poem's finished draft is, the manu-script drafts reveal the stepping stones of progress in the mind of the poet, who appears to have calibrated his belief side-by-side with each draft of the poem. For example, the title change (from "No Pantheist" to "Nothing Is Far") substitutes a more ambiguous stance for an outright declaration of religious orientation. In one draft, the line "What may a common stone conceal?" was originally written "All the believers feel, I feel," then "What the believer feels, I feel," and the word "much" appears to have been crossed out and re-written, as if battling for supremacy with the word "all." Also, the final line "Between the known and the unknown" originally appeared as the more facile and perhaps absurd "Knowing the known, not the unknown." Taken together, these alterations literally depict the poet working through his spiritual orientation, one draft at a time. In doing so, Francis moves from obsessive declaration to tolerant interrogation, from answer to question, and from definitive knowns and unknowns to a comfortable resting point between extremes.

In assuming this non-stance, Francis achieves a deft conflation of the *kataphatic* (or "positive way") and *apophatic* (or "negative way") belief systems. Of these contrary but companion modes of meditation on creation, John Gatta writes that the *kataphatic* mode of belief "encourages engagement with words and images as a means of approaching union with the Creator. . . . By contrast, *apophatic* mysticism involves the systematic negation of all words, images, or concepts of the Divine. Insofar as God transcends all sense knowledge and human representations, the apophatic way leads toward pure contemplation—toward the soul's immediate, unverbalized apprehension of the Absolute" (Gatta, 44). In "Nothing Is Far," Francis combines a superflux of negatives ("never," "not," "no," "nothing") with a flurry of affirmatives and natural images ("all," "clear," "here," "near," "stone," "bird," "tree"), thus achieving a resolution of both "positive-" and "negative-" way avenues of comprehending divinity in nature.

In poetry and practice, Francis demonstrated a positive-negative capability in the way he reveled in a state of being between absolute belief and unconditional doubt. In his autobiography's sixth chapter, "Peace at Fort Juniper," he devotes several detailed paragraphs to the exhumation of a common "jutting stone in the middle of a grassy plot" that he removed to make the place a "bit smoother," unaware that it would become a totemic guardian. "It proved to be an oval face or head," he recalls, "with one squinting eye and a puckered mouth. Nothing was needed but to find a larger flat-topped stone to serve as a pedestal, and behold, I had an idol or rural deity to guard and grace my property." For years, Francis remembers, visitors to his home commented on the stone's true essence: some said "an old man, an old woman, and still others a child." Characteristically, however, Francis remained open to inspiration. "I was more interested in gathering a variety of interpretations," he records, "than in giving my own." This stone god that balanced for "months unmoved" under a choir of birches north of his property remains immortalized in a photograph in his autobiography next to a description that both negates and affirms its spiritual identity (*Trouble with Francis*, 56–57). In a journal entry (July 7, 1953), Francis notes the power of his country stone deity to bring forth authentic and meaningful personal revelation. Briefly, he records a moment of harmonious realization, one that reconciled his father's faith and his path of spiritual divergence. "Sitting in my grove near my stone man this morning, I thought of my father's name: Ebenezer," he recounts. "Why had I not felt it more and incorporated it in my own life? Stone of Help. And by a later meaning, Dissenter's Chapel. Didn't I know the helpfulness of a stone? And wasn't my Fort Juniper a sort of Dissenter's Chapel?" (*Travelling in Amherst*, 88). Permanently enshrined on a bookcase shelf at Fort Juniper today, Francis's modest "eolith palladium" waits to bless and puzzle curious seekers who cross the threshold.

Throughout Francis's journals, he scatters bread crumbs of thought that show how erratically and steadily the pendulum of his belief swung back and forth. Examine, for instance, the seven commandments of spiritual deferment and noncommittal consideration he assembles on July 22, 1931. "How comprehensive and various is the realm of unbelief!" he exclaims, following with a list of items, each beginning with "What I" and including "distinctly disbelieve"; "regard as more improbable than probable"; "think is, at present at least, unknowable"; "think is knowable but on which I have suspended judgment"; "accept as true . . . in a certain sense, not in the generally accepted sense"; "am too little interested in to hold any opinion about"; and "am merely ignorant of" (*Travelling in Amherst*, 5). In another, he slides into the philosophic pews of tolerance and acceptance, seeing the paradoxical benefit of evil. "I see all the evil of the world as a black tapestry against which and within which man is to weave the golden threads of his life," he declares. "He can never blot out the black. He can only weave a more and more beautiful pattern. If he could blot out the black, where would this pattern be?" (*Travelling in Amherst*, 27). Elsewhere, he notes simply, "Religion: hide and seek" (*Travelling in Amherst*, 34).

To sum up Francis's spiritual beliefs becomes, if not impossible, certainly problematic and perhaps unnecessary. At times, he appears to decry religion; at others, he embraces it. "I regard myself as a religious man," he tells Philip Tetreault and Kathy Sewalk-Karcher, "for three reasons": (1) "If religion is essentially one's total attitude toward life, then I have such a total attitude"; (2) "I share and try to follow the same ethical ideals as does the avowed Christian, without accepting the supernatural substructure"; and (3) "though I disagree fundamentally with the religious views of most other people, I am endlessly interested in what they believe" (*Pot Shots*, 126). In synchronic arrangement, three prose sections assemble a loose corral around his elusive, chameleonic pathway to spiritual enlightenment and embattlement: chapter 12 ("A Religion of One's Own") in *Traveling in Concord*; chapter 19 ("The Philosophic Pessimist") in *The Trouble with Francis*; and chapter 5 ("Four Evidences against Theism") in *The Trouble with God*.

"A Religion of One's Own," which takes the poem "Nothing Is Far" as its postscript, addresses the reader directly, someone Francis assumes has reached the unspecified "age of thinking for yourself." As a supplement to exchanging, holding to, or abandoning one's inherited religious faith, Francis offers his readers a fourth option: "an original and modest religion of your own." While he devotes most of his time in this chapter to debunking the dogmas and intellectual dead-ends of the world's religious institutions, Francis takes care to characterize the ongoing evolution of his personal belief in geocentric and ecosystemic terms. "A religion needs channels as a river needs banks," he affirms (*Traveling in Concord,* 131). In contrasting the "great mountain peaks of faith" established by most world religions and the free-thinking individual's "mole hill" of self-generated belief, Francis does not shrink in defense of his more individual credo. "At any moment, I am standing on a very small area of the earth," he writes. "The fact that the remainder of the earth, . . . its mountains and plains and seas, supports me only indirectly—this fact does not undermine my confidence in the spot on which I am now standing. . . . Similarly a religion of my own is of more import to me than are all the great religions, if my religion, like the plot on which I stand, can best sustain my life" (*Traveling in Concord,* 124).

Ultimately, an inclusive biocentrism provides the bread and water of Francis's faith. "In every possible way," he concludes, "to make life-fostering forces predominate in my own life and in the lives of others— this is my religion" (*Traveling in Concord,* 133). As someone who lived a robust and largely independent life into his eighties, he appears to have been abundantly sustained by his faith.

In the final chapter of *The Trouble with Francis,* Francis locks philosophical horns with the nature of evil. In the opening of this chapter, he classifies his "direct, undoctored, natural outlook" on the universe as something that developed after a "long evolution" of personal experience, thought, and struggle. Paradoxically, he labels himself a "pessimist whose personal life had on the whole been fortunate and happy," and he wonders out loud why, when Emerson, Whitman, and Thoreau influenced him so greatly, he hadn't become an "optimist like them?" (*Trouble with Francis,* 223). After peeling apart the connotative fibers of the notion of evil and differentiating between types of evil, Francis focuses on what he calls "evil (3)" or "E": "All that is hostile to human life and its fulfillment" (*Trouble with Francis,* 224). So influential was this chapter on Charles Sides that, in the process of completing his thesis on Francis, Sides referred to it in a letter to Francis following an incident in which a "young boy darted into the path of [Sides's] car and was struck down." Sides elaborates, "I was sure that I had experienced 'E' in its random and destructive manner. Seeing the situation through your poetry helped me resolve the incident in my mind" (Sides, letter to Francis). The bulk of this manifesto Francis devotes to discussing the "fantastic . . . variety of agencies through which E operates": torture, war, natural disasters, accidents, poisonous reptiles and insects, and killer mammals and fish. "If man has gone to excess in his deadly inventions," he queries, "has he not followed the example of nature?" Later, as a result of this line of thought, he adds, "an all-powerful torturing god is not much more plausible than an all-powerful loving god" (*Trouble with Francis,* 232). From the "opposite pole of the orthodox Christian," Francis settles on the pragmatic platform of a visionary reality. "Nature is cruel and ruthless, tender and loving, and utterly indifferent, all at once. Nature includes man himself—how could anyone deny it?" (*Trouble with Francis*, 235). As if to contradict himself with Whitmanesque satisfaction, he concludes, "Though my view of the human situation is dark, in my own life I embrace all available brightness. . . . Love is as precious to me as it would be if there were a god of love presiding over it" (*Trouble with Francis*, 236). Here, Francis's ongoing actions within and reactions to nature as a holy presence appear to agree with what John Gatta calls "our reaction to nature as divine Creation," namely, "something both authentically discovered, or discoverable, *and* humanly constructed" (Gatta, 10).

"Precisely where does the work of God leave off and the work of Nature begin?" Francis begins *The Trouble with God*, a book that Robert Shaw accurately labels a tad "weary and rambling" (*Trouble with God*, 1; Shaw, 88). *The Trouble with God*, while certainly not Francis's most engaging prose work, sets its hooks in some tough questions and centers its debate in the mingled regions of human and non-human interrelationships. "How could an all-wise all-loving creator have devised an order of nature on earth in which every species of animal is the necessary food of other species," he challenges, "so that to sustain its own life a creature must constantly destroy other lives?" (*Trouble with Francis*, 54). Unflinchingly, Francis levels the crosshairs of his theological cannon at those who "accept nature's dog-eat-dog process" as part of nature's drive toward a "mystic balance" so that "the health of the earth's biosphere is maintained." And yet, "under the same law," Francis questions the supposed benignity of nature when human beings find themselves "reduced in numbers . . . by flood, famine, epidemic, earthquake, and war." Even Thoreau does not escape the vise-grip of Francis's nutcracker skepticism. "Henry Thoreau, an exceptionally thoughtful man," he notes, "was never able to or willing to see that what he castigated in human society was no other than what he blandly accepted and even praised in nature" (*Trouble with God*, 55–56). Though it adds a useful dimension to the characterization of Francis's unique brand of eco-spirituality, and an understanding toward the roots of his skepticism, the trouble with *The Trouble with God* is that it repeats much of what Francis wrote elsewhere, and it lacks the vitality, splendor, and authorial candor and vulnerability that readers encounter in his earlier works.

In his landmark article "The Historical Roots of our Ecological Crisis," Lynn White Jr. boldly and memorably called for St. Francis of Assisi—based on St. Francis's radical insistence that humans should live inseparably from nature—to be universally ordained the "patron saint of ecologists" (White 14). Decades previous to the publication of White's article, a wayfaring Robert Francis, while in Rome, attended an "extraordinary ceremony" celebrating the seven-hundredth anniversary of the death of Brother Juniper, "the most amusing and possibly most saintly of the first followers of St. Francis." Though his fellow artists at the American Academy showed little or no interest when he told them about his view of the ceremony, Francis thrilled at witnessing the ceremony firsthand: "the box of bones, looking like an oversized box of chocolates tied with a ribbon, deposited in a niche inside a pilaster next to the high altar" (*Trouble with Francis*, 113). Given the similarity of their names and passion for embedding their lives within the rhythms and patterns of nature, it is possible to read into history a certain kinship between the two Francises—saints separated by centuries but united in philosophy—though, according to Henry Lyman, when people in Rome asked Francis his name, he would reply, "Franciso. Ma non santo"— "Francis, but not holy" (Lyman, e-mail to author, August 28, 2008). In "Manifesto of the Simple," from his posthumous collection *Late Fire, Late Snow*, Francis refers to St. Francis's influence in his closing lines: "Who said Be simple? Jesus. / Who said it again? St. Francis. / . . . We are Mother Earth's simple ones. / We are Mother Earth's simpletons" (13–14, 16–17). Though Francis did not hide his critiques of—and, at times, contempt for—organized religions, he was a man of remarkable spiritual sensibilities and a believer in the higher powers of a universal nature. In a way, it could be said that St. Francis and Brother Juniper enjoyed a reincarnation in Bob Francis of Fort Juniper.

If St. Francis of Assisi qualifies as the patron saint of ecologists, I would argue for Robert Francis to be seen as the patron ecopoet of the uncommon experience. After the publication of Francis's *Like Ghosts of Eagles* in 1974, Andrew Salkey, in a BBC press release, affirmed Francis's distinctive spiritual capacity. "If the finest poets may aspire to the status of prophets," Salkey observed, "then Robert Francis's best poetry as prophecy is a rock we can all depend on, I believe" (Salkey). Fran Quinn recounts that after accompanying his students on a field trip to visit Francis at Fort Juniper and the Trappist monastery in Spencer, Massachusetts, his students considered Francis's lifestyle "holier" than that of the monks, whom the students saw as living an opulent lifestyle (Quinn, interview with author, December 8, 2008). Given Francis's background, and the spiritual ground over which he traveled and which he sedulously turned over with the rusty spade of his thoughts and words, it is not difficult to read a trace of autobiography in his mini-essay "Neither":

> I remember so well how he looked when he said it and how they looked when they heard it. They had asked him the old chestnut.
> "Do you write for yourself or for an audience?"
> "Neither," he answered mildly.
> Sputtering surprise. "Well, who—who *do* you write for?"
> "I write for God."

THE STATE OF NATURE

"War never won me over," Francis opens chapter 4, "The War," in his autobiography. "The waving of flags and the blowing of bugles never dazzled or seduced me" (*Trouble with Francis,* 32). Francis's staunch anti-war stance derived in part from his personal commitment to pacifism and his lifelong pilgrimage toward promoting concinnity between humanity and surrounding biospheric elements. Throughout his poetry and prose, Francis consistently pursues his belief that to flourish in one's natural settings one must also oppose the politically motivated destruction of wars waged against the earth's human populations and against the progress and rejuvenation of the earth's ecosystemic processes. Francis's twin devotions to the preservation of life and landscape cast him as an American writer who saw no division between the political and the ecological, a position with which a generation of ecocritics would come to agree.[5] Jonathan Bate, in *The Song of the Earth,* acknowledges this viewpoint. "The dilemma of Green reading," Bate argues, "is that it must, yet it cannot, separate ecopoetics from ecopolitics" (Bate, 266). As someone who lived through World War I, World War II, the Korean War, and Vietnam, Francis sidestepped the decades of international conflict and armed himself for combat with pencil and typewriter. Opposed to violence, he hummed the battle hymn of the hummingbird, joined the ranks of the soldier beetle, pledged his allegiance to peace, and aimed his most explosive poems at the wastefulness of twentieth-century warfare.

Writers and critics have often highlighted Francis's gift for deftness and subtlety, but in his most volatile anti-war poetry, he exchanges deft subtlety for deafening subversion. "Bruised by American foreign policy," the speaker in "Hogwash" laments, "What shall I soothe me, what defend me with / But a handful of clean unmistakable words" (7–9). The experimental post-Vietnam "Poppycock" crows, "ballyhoo from Madison A / ballyhoo from Washington DC / red-white-and-blue poppycock / Hurrah!" (21–24). The single-sentence satire "I Am Not Flattered" critiques those who tolerate wars from a distance: "I am not flattered that a bell / About the neck of a peaceful cow / Should be more damning to my ear / Than all the bombing planes of hell / Merely because the bell is near, / Merely because the bell is now, / The bombs too far away to hear" (1–7). Perhaps autobiographical, "The Articles of War" recounts an anonymous twentieth-century soldier's simplistic desire to "resign from the Army," a question which swells his barracks with derisive laughter (9, 13). "Somebody aghast at history," the speaker finishes, "may try / Resigning from the human race / . . . Haunted by the hawk's eyes in the human face / Somebody—could it be I?" (16–20). "Light Casualties" satirizes the sanitized politically correct phrase it takes for its title: "Did the guns whisper when they spoke / That day? Did death tiptoe his business?" (7–8). The fragmented and experimental "Blood Stains" lifts a home remedy for "how to remove" blood from household fabrics and applies it to the global need to cleanse the blood of war from "headlines dispatches communiqués history / white leaves green leaves from grass growing / or dead from trees from flower from sky / from standing from running water" (9–12). When it came to global conflict, Francis favored a direct charge in place of a flanking maneuver.

At times, the shocking violence of war is addressed through Francis's jarring syntax. One grammatically relaxed piece, "The Righteous," damns wars fought under the guise of religion, its disconnected lines streaming like an AK-47 sermon:

> After the saturation bombing divine
> worship after the fragmentation shells
> the organ prelude the robed choir after
> defoliation Easter morning the white
> gloves the white lilies after the napalm
> Father Son and Holy Ghost Amen.

In a similar spirit, "(Two Poems)" connects U. S. military gasconade with idolatrous Old Testament gods: "where is the public man / year after year / unstainable / . . . Do not bend / do not bend the knee / to Baal / to Moloch / to the Pentagon / . . . do not fold the hands" (3–5, 8–14). In "Time and the Sergeant," the former Army private turned poet borrows the dark offhand-edness of e. e. cummings's "Buffalo Bill's" and asks the ghost of his former drill sergeant, "Has Old Bastard Time touched / Even you, Sergeant, / Even you?" (13–15). And finally, still indefatigable in his eighties, Francis composes "Varieties of War," a sardonic catalogue of modes that includes "shooting war, non-shooting war," "pushbutton plushbottom," "who the hell are we fighting?," "War to make the world safe for democracy / War to make the world safe for dictatorship / The President's war on Poverty," and "The President's war on poverty-stricken Asiatics" (5–6, 16–20). Depicting war as a shameful commodity in a capitalistic world, Francis's persona concludes, "If you can't find it at your supermarket / Look, Buddy, what in hell do you want anyhow?" (23–24). With blitzkrieg bluntness, Francis dug in at Fort Juniper for over four decades and from his biocentrist bivouac launched a full-frontal assault at international conflict.

Though he favored direct satire in his anti-war diatribes, Francis also authored a significant number of ecopoems in which he indirectly attacked the United States' military and political muscle by deconstructing its most recognizable icon—the American eagle—as an appropriated ecological symbol of both power and predation, danger and endangerment, freedom and fury. In "Eagle Caged," Francis's speaker demotes

America's national raptor and renders it "Uneagled" in a "coop" like a "feather duster" (1, 4). In the philosophically candid "Eagle Plain," the speaker simply states that the "American eagle is not aware he is / the American eagle," thus emphasizing how human culture has inappropriately and vainly appropriated a wild bird to advance a national agenda (1–2). In a country based on freedom and independence, the speaker adds, its central symbol remains free and independent: "The American eagle never says he will serve if drafted, will dutifully serve etc. He is / not at our service" (10–12). At poem's conclusion, the speaker shows how Americans have failed to capture the image that has captivated them with a national ethos of valor: "If we have honored him we have honored one / who unequivocally honors himself by / overlooking us. / He does not know the meaning of magnificent. / Perhaps we do not altogether either / who cannot touch him" (13–18).

In "The Big Tent," the American eagle is reduced to a cliché circus act that clutches in his claws an "iron olive branch" and perches next to a mad parrot that shrieks, "Hells bells! E pluribus unum!" (11, 16). "The Disengaging Eagle" depicts the eagle, according to a "rumor," finding "his official pose / faintly absurd" and aspiring "to unofficial peace" where he can be "pure bird" (1, 8–12). The resigned and distant speaker expands the elaborate rumor by claiming that the American eagle wants to "abdicate / forever and for good / as flagpole sitter for the State" and that "the old warrior / who screamed against the sun / and toured with Caesar and Napoleon / cavils now at war" and wishes to "retreat / to blue solitude" (14–16, 23–29). And lastly, in the prose poem "Eagle," Francis contrasts the eagle's natural characteristics to the rapacious qualities of a country that has, in application, misinterpreted the metaphor it adopted. "Strong bird," the speaker opens, "if we could have your strength without your violence," "your unfailing flight without your crooked hands," "if strong were not aloof," "if free were more than unconfined / . . . you would be stamped on something better than our gold" (1–5). As a subset of his pacifist and anti-war lyrics, Francis's eagle ecopoems take flight from fighting and re-direct the American gaze toward the contradictory political and environmental ideologies screaming for release from its national symbology.

In his *Pot Shots* interview with Tetreault and Sewalk-Karcher, Francis sheds additional light on his anti-war poems. Francis ranks many of the poems examined here as anti-war but not "anti-war propaganda." "Propaganda aims to incite action; poetry leads to reflection," he contends and then, surprisingly but candidly, claims he failed in using poetry to change people's "attitudes and actions" concerning war. In the same interview, Francis reveals how, perhaps as a reaction to his feelings of having failed to generate successful Amherstian agitprop, he stood "with a group of

Sunday vigilers" on Amherst Common from noon to 1:00 p.m. to pro-
test the Vietnam War, something he did regularly from 1966 to 1973
(*Pot Shots*, 140). Elsewhere in the *Pot Shots* collection, he authors sev-
eral miniature experimental essays that portray the many shades of his
pacifism. "Caught in a Corner" defines an "antipoem" as a "bullet or a
puff of poison gas" (*Pot Shots,* 159). "Francis" consists of three double-
spaced lines, two complete, the final line incomplete: "Francis Scott Key:
'The Star-Spangled Banner.' / Samuel Francis Smith: 'My Country 'Tis
of Thee.' / Robert Churchill Francis: " (*Pot Shots,* 188). Here, the blank
after the third and final colon remains weighted with Francis's sentiments
toward American sentimentalism. In other words, he has written himself
out of a discourse of sentimentalism by not writing himself in. Of all the
essays in *Pot Shots*, "Poetry as an Un-American Activity" remains engag-
ing because it consists of a single dependent clause that while grammati-
cally unfinished, distances poetry from politics and ends with a central
ecological image:

> If the notion ever got around that poetry was an un-American activity
> (as in some respects it certainly is) and that poets were dangerous people
> (as some of them surely are), and if instead of being encouraged with
> prizes, awards, gold medals, fellowships, and other subsidies, they were
> penalized and even suppressed—
> So that to escape fines and prison sentences and loss of good name
> and employment as college teachers, most poets stopped writing poetry
> altogether—Except when, on very rare occasion, a poem so insisted on
> being written that the poet yielded to temptation despite all risks and
> wrote a poem he was willing to die for—
> If, in other words, poetry almost disappeared from the earth and only
> occasionally in some out-of-the way place bubbled up like a pure moun-
> tain spring—(*Pot Shots*, 48)

As seen here, Francis, through humor and dark sobriety, wrote to dis-
tinguish between the idea of America and American ideals, as well as
political and ecological America.

In his journal and prose reminiscences, Francis sketches out the saga
of his development as a conscientious objector to war and protector of
the environment. "It was during my freshman year in high school that I
first went on the record against war," he remembers. In several places, he
recalls a few patriotic actions and decisions that stemmed from his inner
feelings of being called to "Duty": a summer month at Camp Devens
under the "Citizens Military Training Camp" and his registration for the
Selective Service in 1942 at the age of forty-one (*Trouble with Fran-
cis*, 32). "I am liable to the draft," he notes. "I am faced with a conflict
of duties: duty to my country, and duty to my integrity. I am prepared

to suffer for my country, but not to cause suffering for it" (*Travelling in Amherst*, 55). The year his conscience turned one hundred eighty degrees, "Duty was pointing in the opposite direction," and he joined the War Resisters League, whose pledge he includes verbatim: "War is a crime against humanity. I therefore am determined not to support any kind of war international or civil, and to strive for the removal of all the causes of war" (*Trouble with Francis*, 32).

Chapter 4 in his autobiography, "The War," presents a detailed and enlightening account of Francis's short-term, bottom-rung existence as an Army private, complete with accounts of his intestinal troubles and charley horses in his feet during basic training. As a means of steeling himself against his participation in what he believed to be a corrupt and violent military enterprise, he conjured up the presence of his literary forbears. While marching and drilling, he recalls, he imagined "invisible marchers" beside him: "Thoreau with his long steady straight-ahead stride" and "Whitman with a more leisurely gait, Whitman who had never been a soldier . . . but who had been close to soldiers and to war" (*Trouble with Francis*, 42). In his journals, he records this episode with added details. While marching, he imagined Thoreau always "half a step ahead" of him and wondered if it was Emerson who saw "something military" in Thoreau's manner. Socrates, too, appeared alongside Francis at Camp Breckenridge, "sturdy and untiring." "What made him so welcome a companion," Francis writes, "was knowing that he too had been a soldier and a good one, and that he could be a soldier and Socrates at the same time" (*Travelling in Amherst*, 56). With pride, Francis includes a letter he wrote in 1947 in which he declines an invitation to contribute to the Harvard Class of 1923 gift fund, his reason being that his six-hundred dollar annual income allowed him just enough "to send a little food now and then to a family in Germany" (*Trouble with Francis*, 218). With more practicality than pride, he also mentions that he donated a portion of his pauper's income to help a local group of Quakers "take money to Canada to purchase medical supplies for Vietnam south and north" (*Trouble with Francis*, 220). Though he emphasizes that he was "honorably discharged," Francis describes the end of his enlistment as a welcome abandonment: "My country was now in a position to permit me to go on with my own life—in short, to forget me" (*Trouble with Francis*, 54). Without apology, he purposefully mentions that, thanks to the U.S. Navy's purchase of a thousand copies of his collection *The Sound I Listened For*, it enjoyed a second printing, though he remarks, "My poems went to war to an extent that I never did" (*Trouble with Francis*, 77).

In November 1960, the year *The Orb Weaver* was published, Francis visited Syracuse University for a three-day stint as a guest author. One historical anecdote related to his brief tenure there brims with subtle satire concerning his feelings about poetry and politics. "I was there the day Kennedy was elected," he remembers. "Unsuspectingly I had slept above a polling booth. When I came downstairs in the morning, having been awakened by an unaccountable bustle below, voting was already in progress. Poetry upstairs and politics down: this was the note on which I began my reading that afternoon" (*Trouble with Francis*, 121). Unlike Socrates, Francis never became a good soldier, according to U.S. military standards, though he served his time. Rather, from the common man's campanile of Fort Juniper, he crusaded for tolerance and peace. His poetry bore record of one man's battle to understand the politics of ecology and to reconcile a combative humanity with its organic allies.

6
Economy, Place, and Space

The frog does not drink up the pond in which he lives.

—Lakota proverb

To the creator there is no poverty and no poor indifferent place.

—Rilke

IF PACIFISM AND POLITICAL ACTIVISM CHARACTERIZE Francis's twentieth-century American ecopoetics, so does a concern with economics, both the human-made and organic varieties. It is not too much of a stretch to label Francis a pauper-poet, nor to see him as a steward of places, since he "stayed put," as Scott Russell Sanders has advocated, and since he subsisted on so little for the majority of his life. This poverty, however, served as both crisis and credo. In material terms, he consumed very little energy, natural resources, and food—just enough, and sometimes not nearly enough, to live on. As a result, he authored a slim body of prose and poetry whose chief characteristic is its sparing use of language. In most cases, it is this single technical dimension—the use of few words—that gives his pieces their forcefulness and remarkable clarity. While Francis's economic straits strained his soul, they also provided him a kind of enrichment and, eventually, a full-fledged aesthetic all his own. This aesthetic blends human economics, the economy of the natural world, a commitment to the conservative use of language, and a devotion to what ecocritics have referred to as "place-making"—treating the places of the earth as vital resources and endowing natural settings with human value.

THE AESTHETICS OF ASCETICISM

In February 1967, a twenty-five-year-old Paul Theroux sent Francis a letter from Kampala to share with Francis the news of the publication of his first novel, *Waldo*. Surviving historical sources indicate that Theroux, while attending UMass as an undergraduate, formed a friendship with Francis around the time that Francis turned sixty and began a modest but noticeable resurgence into the national literary scene.

Theroux's association with Francis appears to have stemmed from their shared anti-establishment political sensibilities and passion for writing. In the opening of his letter, Theroux berates UMass for naming its newest dormitories after John F. Kennedy and John Adams—"our finest fascist"—and lists a selection of more desirable choices, including Thomas Paine, Eugene V. Debs, and A. J. Muste, though "Juniper Hall" playfully tops Theroux's catalog of suggestions. Having flitted through politics and literary successes, Theroux turns to the subject of financial security. "By the way, I am looking for a job," he reports. "I know you are the wrong one to ask, having stayed prosperously and cheerfully unemployed since you left Beirut. . . . R. Francis has never taken a job and yet looks the world in the eye. What would it cost me to build a house like yours? Does it take money or guts?" (Theroux to Francis, February 13, 1967).

Though Theroux exaggerates the facts about Francis's employment history, he touches on a definitive dual aspect of Francis's life and artistic motifs, one that surfaces repeatedly in Francis's unpublished and published poetry and prose: the "economic question," as Francis calls it (*Trouble with Francis*, 215). Between Thoreau, whose extensive "Economy" chapter invokes *Walden*, and contemporary ecocritics who have mulled over the etymological significance of the environmental concept of "economy," Francis's twentieth-century ecopoetry remains intriguing because it dwells with unflinching candor on the subject of the environmental and artistic benefits of self-elected poverty while exhibiting a technical fidelity to the most economic use of language possible.[1] In short, from a conservationist standpoint, Francis's poems are *about* but also *demonstrate* economy.

First, I turn to those poems in which Francis mulls over the economic characteristics of ecological phenomena, poems in which he takes for his express subject the naturally generated, calibrated, and maintained economy of nature—as well as humanity's proper relationship to nature's economy. In an early poem, "Homeward," a piece that inspired another David Leisner classical guitar composition, Francis's speaker meditates on the possibility of returning to the soil, of making his house, so to speak, directly in the material of the earth: "Sun that gives the world its color, / Turn me darker, deeper, duller. / Make the clouds white, and the foam. / Make me brown as fresh-turned loam. / Save whiteness for sky and sea. / Give the tan of earth to me. / Blend me with the hue of loam. / Turn me homeward, turn me home" (1–8). Through energetic structural brevity, "Homeward" captures readers with its message of stark suggestiveness: that the ideal state of human nature is to live closer to the soil, perhaps ultimately mingled within it, atom for atom.

The etymological connection that joins matters of ecology and economy—the *oikos* that Lawrence Buell refers to—operates here on a subliminal level, addressing Wendell Berry's transformative and holistically regenerative dialectic of the human "little" economy and the ecocentric, ethical, and spiritual "great" economy. In a personal meeting with Robert Frost (March 1933), Francis listened while Frost critiqued the artistic merit of "Homeward" and other poems in Francis's first manuscript collection. "Homeward," Francis remembers Frost saying, was singled out for "honorable mention," and Francis recalls that Frost praised his poems generally for "being not too long." In this same meeting, Frost complained to Francis of the way that "many modern poets" extended their poetry to "undue lengths" (*Frost: A Time,* 50). Frost's observation would prove prescient in that, throughout his life, Francis pursued the study of the earth's economic properties through a scrupulous and unflagging dedication to avoiding the waste of his most precious resource: words.

In a variety of ways, Francis managed his household economy (both physical and psychological) in a manner that reflected the management of the world's naturally maintained economy. His poetry, sometimes as an expression of harmony or discord between the two, acts as a record of this personal and environmental management. In "Poverty Grass," the speaker delights in a sweeping panorama of "field after field of poverty grass" that he spots beyond "the tenements of the town / Beyond the tin-roofed hoi polloi, / The railroad and the highway pass" (1–3). "Rare is the passerby that knows it / Or knows the irony of its name," the speaker concludes (16–18). In this declaration, Francis implements a moment of ecodialectics and reverses a social outlook that sees humans as rich and organic elements as poor, specifically the otherwise ordinary poverty grass, "which, heaven knows, is rich enough / In every hue from red to brown" (5–6). Though Francis doesn't stipulate the source of his inspiration for writing "Poverty Grass," a footnote in *Frost: A Time to Talk* preserves an exchange between Frost and Francis, a moment when Frost is forced to ask Francis the name of the grass, after which Francis, the self-schooled naturalist, provides the Latin name: *Andropogon scoparius* (*Frost: A Time,* 18). Thus the humble ecopoet proves himself poor in finances, but rich in substance.

Intuitive principles of preservationism operate throughout Francis's poetry collections, gradually inviting readers to undergo a conversion to an ongoing conversation concerning more economical ways of inhabiting the earth. Unwilling to waste a molecule of time, aroma, sound, energy, or space, the speaker in the trite but memorable "Salt" savors his devotion to conservation: "Salt for taste / And for wit. / Be wise. Don't waste / A pinch of it" (11–12). Elsewhere, in "The Stones in My Life," an uncollected *Christian Science Monitor* essay, Francis meditates on the small "stones of economy" that he would warm up and hold in his gloved hands against the steering wheel as he drove his unheated car in January. "Though they do not stay hot indefinitely," he philosophizes, "they are warm long enough for my purpose. And the heat costs me nothing, since all I have to do is to lay the stones on the stove a few minutes before I go out" ("Stones" 8). While Francis struggled mightily with poverty, he rarely assumed a tone that did not thankfully celebrate the bountiful wealth of the earth's implements.

From a stance that might seem fatalistic to some but conceptually sublime to others, Francis depicts the earth as a perfectly regulated recycling organism whose "housekeeping" modes and processes eventually school human inhabitants in the realities of universal economics and conservation. "A Health to Earth," which runs its title into the body of the poem, knits together loosely jotted tercets to communicate this theme: "nothing defeats her nothing escapes / the owl ejects an indigestible / pellet earth ejects nothing / she who can masticate a mountain" (3–6). To the speaker, the households of human and non-human life forms eventually combine in a sedulous display of energy conservation: "all man's perdurable fabrications / his structural steel, his factories, forts / his moon machines she will in time / like a great summer-pasture cow / digest in time assimilate / it all to pure geology" (10–15). Thus, the earth's slow progress supersedes humanity's rapid rise and fall.

In "Cold," from the perspective of the subject trapped mid-winter in impoverished circumstances, Francis's speaker describes a single-person household and how the occupant survives by reducing his personal energies at a time when the earth's external household appears to be doing the same. The barbed texture of the poet's language mimics the frigid snap of a New England frost:

Cold and the colors of cold: mineral, shell,
And burning blue. The sky is on fire with blue
And the wind keeps ringing, ringing the fire bell.

I am caught up into a chill as high
As creaking glaciers and powder-plumed peaks
And the absolutes of interstellar sky.

Abstract, impersonal, metaphysical, pure,
This dazzling art derides me. How should warm breath
Dare to exist? Exist, exult, endure?

Hums in my ear the old Ur-Father of freeze
And burn, that pre-post-Christian Fellow before
And after all myths and demonologies.

Under the glaring and sardonic sun,
Behind the icicles and double glass
I huddle, hoard, hold out, hold on, hold on.

In the mingling of human breath and atmospheric fire, the clash of icicle consonant and storm window syllable, Francis communicates the perpetually unfinished phenomena of *environmental isolation* and *insulation*. This paradoxical moment finds the human subject both preserved and threatened, encased and liberated. In Francis's case, his surroundings become a half-natural/half-constructed but interdependent combination of wild and human-made "households." While the restless poet strains against the trap of his human confines, he would be harmed if he were to leave their asylum. The poet's *pis aller* becomes the act of writing as refuge. Crisp rows of c's collect on the first stanza like icicles on raingutters and harden in the subzero breath of the reader. The susurrus of repeated e's and h's intermingles the exhalations of chilled air and indoor outlook of the poet. A sparse arrangement of solid-ice lines and words invites readers to enter the severe outdoor habitation of a New England winter while remaining sheltered in reading.

"As to language," Richard Wilbur writes in the introduction to Francis's *Butter Hill*, "New England has always relished economy" (Wilbur). In proffering this assessment, Wilbur touches on a second dimension to Francis's passion for conservation: the sparing use of language. For Francis, this tendency to not waste words functions as an aesthetic outgrowth of his world view, one that derives from nature's tendency toward conservation and recycling. His sparse poems, many of which reach a maximum of eight lines, serve as lexical reminders to readers. Physically, as textual symbols, they embody the ideal environmentalist commitment to producing much by expending and consuming little.

In "Before," for example, the poet overturns egocentric versions of the Creation and condenses nature's immortal supremacy over human sensory intelligence into three couplets: "The sea was blue, the sea was green / Before the sea was ever seen. / Surf muttered its liquid word / Before the surf was ever heard. / And waves made time against the shore / Before mind thought its first Before." In the beginning, Francis suggests, in an effort to rewrite religious history, was the world—and *then* came the word. "Enigma," another chiseled gem from his earlier books, erodes an entire epoch of world culture in six lines: "Nothing Egypt did / In her dark pyramid / Remains forever hid. / The undeciphered land / Is here beside my hand—/ This pyramid of sand." Similar to "Before,"

which shows how oceanic currents pre-date the earth's human pres-
ences, "Enigma" argues that the persistent mill of the earth's eroding
agents has outlasted and will always outlast human culture and reduce it
to a memory.

Many of Francis's economical ecopoems depart from vast commen-
taries on world history and culture to focus on the individual's quest
for meaning in partnership with the natural world. The single quintet,
"By Night," collected in William Cole's 1973 anthology *Poems One Line
and Longer*, recounts the fleeting thoughts of a figure awakened by a
scream "[a]fter midnight": "But whether it was bird of prey / Or prey
of bird I could not say. / I never heard that sound by day" (3–5). The
seven-line impromptu "As Easily as Trees" invites a man with cabin fever
to "drop / All rags, ambitions, and regrets" and lie "with leaves in sun"
(4–5). "Blue Jay," whose brevity prompted William Cole to include it in
his 1967 anthology *Eight Lines and Under*, sketches a thumbnail portrait
of the "bandit-eyed" and "undovelike" bird, the "skulker and bluster /
whose every arrival is a raid" and whose presence reminds Francis of his
late father's absence—"Still, still the wild blue feather / brings my mild
father" (1, 3–4, 7–8). Of the conservative energy and stylistics in "Blue
Jay," David Young writes, "The accomplishment is the more impressive
for its economy and concentration. The whole poem is paradoxical by
virtue of those qualities too: a short, small thing that weighs more, means
more, requires more attention, than we thought likely" (David Young,
"Francis and Blue Jay," 309). In other poems, a single word suffices for an
entire line, as is the case in the weather-vane shaped stanzas of "Weath-
ervane": "Moving unmoved, / Like the fixed tree / for constancy / But
like the leaf / Aware / Of all the tricks / And politics / Of air. / Fickle? /
Let the fool laugh / Who fails to see / That only he / Who freely turns
/ Discerns, / Moving unmoved / Is free." The most economical way of
living, Francis's speaker suggests here, is to let the wind's muse turn us
where it wants to.

In addition, Francis's poems of economy display a gamut of tones,
from playful to painful, from scherzo to schizophrenic. At times, his
passion for eliminating needless words produces much-appreciated
humor amid many of his more somber lyrics. In a series of epitaphs,
the vegetarian philosopher composes a nutshell treatise on death for a
"Butcher"—"Falleth the rain, falleth the leaf / The butcher now is one
with beef." The inevitable descent of humanity to the earth's embrace,
however, remains a subject for serious reflection in "The Giver," which,
perhaps inspired by Robert Herrick's "On His Departure Hence,"
remains one of the few poems in English written in iambic monometer:
"He gave / His life / Or so / We say / He gave / His death / But now
/ We know / Here by / His grave / How he / Gave both." Ultimately,

though, Francis's most captivating display of ecopoetic economics must
certainly be "Prescription":

> Whoever would be clean
> Of cluttering desire
> Must scrap the golden mean
> And bed with frost or fire.
>
> Only two ways to cure
> The old itching disease:
> No middle temperature
> But only burn or freeze.

In this succinct eight-line parable, Francis etches a code for living that
encourages a humanity ruled by its own selfish passions to embrace an
asceticism of simple extremes meted out by the surrounding weather and
climate patterns. Burn and freeze as regularly and simply as the earth,
Francis's speaker admonishes readers who would otherwise be ruled by
human desire. As a hot and cold globe revolves between poles, the poet
says, so should those who inhabit that globe.

In Francis's collected *Pot Shots* essays, some of which measure a single
sentence in length, the intertwined themes of poverty's beneficence and
the aesthetic payoffs of linguistic frugality return in his hallmark palette
of tones—somber, self-reflexively satirical, and off-the-wall. In "It Really
Isn't," Francis's alter ego blithely exults that "Homer managed with less"
than what is required of modernist poets: pen, pencil, paper, envelopes,
stamps, and a typewriter. Frankly, the speaker adds, poems written in an
"Italian villa or a French château will be no more immortal than those
written in your bedroom at home." In fact, he insists, the poet doesn't
need the "aid, assistance, subsidy, and support that munificent philan-
thropy stands ready to grant him." In this he exults, "Isn't he lucky?" In
the end, though he received a Rome fellowship himself, Francis argues
that if you give a "free year in Rome" to a poet, "it may turn out that what
you have chiefly done is add to his baggage" (*Pot Shots* 25).

Similarly, in "Poetry and Poverty," Francis plays on the lexical resem-
blance between the two words in his title and meditates on the many
ways that "cottage life" in the country, as opposed to "going hungry in
New York or Paris," could be good for the struggling artist and "being
good for him good for his poetry as well" (*Pot Shots* 81). In the Tet-
reault and Sewalk-Karcher interview, Francis reveals certain biographi-
cal details about his passion for economy. He says, "My mother inspired
my economy far more than did any Eastern philosophy." After cataloging
how his mother enhanced leftover dinners and endlessly remade dresses
and hats, he leaps immediately to a description of how her conservative

economic approach to living directed his poet's aesthetic: "In all this I was doing with my life what my poems were doing as they came into being. A poem in its early stages has a healthy appetite and wants to take into itself every available resource. At the same time it wants to eliminate whatever does not really belong to it. Thus a poem develops and achieves true selfhood" (*Pot Shots*, 122).

In another short essay, "My Life," Francis declares aesthetic war on the cumbrance of a single indefinite article. In this microcomposition the speaker complains that his audiences have often misrepresented the title of his piece, *Rome without Camera*, as *Rome without a Camera*. The invasive addition of an unnecessary vowel, Francis maintains, "made a difference in balance, rhythm, and economy," and he declares himself ready to lay down his life to keep his original title "inviolate" (*Pot Shots*, 158). In the essayette "Two Words," the speaker claims that "two words are the absolute minimum" required to make a poem, his reasoning being that one word "cannot strike sparks from itself" (*Pot Shots*, 149). As if to turn *theoria* to *praxis*, Francis infuses the concept in "Two Words," the essay, into twelve tight lines of ecophilosophy in "Two Words," the poem: "Two words are with me noon and night / Like echoes of the solitude / That is my home—half field, half wood, / Feldeinsamkeit, Waldein-samkeit. / In words as quiet as the Psalms / I hear, I overhear the tone / Of Concord and of Emerson, / And all the autumn mood of Brahms. / However foreign to my tongue / They are familiar to my mind / As is the breathing of the wind / And all the wind has said and sung." The most economical use of one's life, Francis observes, is to purchase a solitude of priceless tranquility, an energy of motionlessness, of "field-aloneness" and "woods-aloneness" in the constantly self-refurbished household of the surrounding wilderness.

In chapter 15 of *Traveling in Concord*, "A Poor Man's Assets," Francis salts his foundational economic philosophies and aesthetic principles with wry humor. In doing so, he stakes out further ideological terra firma on which he argues for the benefits of poverty in the life of the artist subsisting on the gourmet standard fare of nature's bare-minimum banquet. "Why should we pity the poor man? Why need poverty be always grim?" he opens and then points out that, for the ecopoet, "poverty can be as positive as an athlete stripped for action" (*Travelling in Concord*, 158). With the lucid axiomatic candor of a roadside sage uttering proverbs, he mulls over the situation of the thinning and somewhat malnourished poet whose lack of material wealth has nudged him nearer to death: "If a thin man has enough flesh for health, why should he want to be fatter? If he weighs enough to not be blown off the earth in a March wind, why should he want to weigh more? . . . For a poor man death is a relatively simple affair since he has little to lose except his life" (*Traveling in Concord*, 159). In sketching out the artistic benefits of penury, Francis clearly favors a rural environment over urban settings, the city being far less "generous" to the poor man than the country. In addition to the boundless varieties of "edible weeds," Francis lists the greatest of all country gifts as "silence"—a "muted and infinitely varied music to which birds, insects, barnyard voices, rain, and wind contribute their quotas." To the outdoor aesthete, "In the country, all that the ear hears is or could be music. And all that the eye sees could be landscape or genre. Though it sees tragedy, it is the immemorial sort related to stones and oaks and stars" (*Traveling in Concord*, 161).

Chapter 15 of *Traveling in Concord*, in addition to cataloging nature's gifts and thrift, covers the poor man's "makeshifts," specifically the construction of "an outdoor seat," a ramshackle but serviceable implement assembled from two scrap boards painted white and "three old automobile tires too worn to have any trade-in value," an idea the poor man happens on after searching "the environment and his mind" (*Traveling in Concord*, 162). In this brief anecdote on self-reliance and practicality, it is impossible to miss Francis's passion for pinching pennies, let alone his genius for recycling and recreation. His unorthodox car tire usage segues into his final excursus on another "advantage of poverty": the "frequent necessity of walking." He maintains that "any pace faster than a walk is too fast for seeing. Eyes and legs, not eyes and motor cars, evolved harmoniously together" (*Traveling in Concord*, 164). From the experienced standpoint of someone who flourished for years without a car, Francis maintains, "A man is as poor as he feels. The poor man that I have been talking about is, of course, not poor at all" (*Traveling in Concord*, 165). Here, Francis takes the stance that mechanization and its supposed advantages enact clear disadvantages on the artist.

Francis's published journals, too, are replete with nuggets of survivalist economics, moments when he buckled under the realization that his coffers were empty—and when he rejoiced in the weightlessness of that emptiness. Richard Gillman, in the introduction to the journals, emphasizes that Francis, in his post-Harvard thirties, "began learning how to economize" in "preparation for a life that would almost always be lived at a near-subsistence level," but a life that was nevertheless a "natural response to his personal priorities" (Gillman, Intro. to *Travelling in Amherst*, xi). On December 1, 1931, Francis writes, "A bill for money owed me lies on my desk. I cannot send it now because it requires a two-cent stamp and I have exactly one cent in my cash box" (*Travelling in Amherst*, 9). Over a year later, in March 1932, he exults when *Virginia Quarterly Review* publishes his two "Dark Sonnets," and he jokes sardonically, "Now I can *afford* to starve." He then records a moment of reflective aporia as he considers the exchange value of poetry in terms of the commercial value of natural resources: "Yesterday afternoon two cords of wood were put into our cellar. Two cords of wood, price fifteen dollars. Two sonnets, price fourteen dollars. I stood in the cellar looking at the small mountain of wood, and thinking" (*Travelling in Amherst*, 10). On April 7, 1933, he lists his "determining reason" for boarding at the home of Margaret Hopkins as "economic"; ten days later, he sighs, "What shall I hang on my walls (when I have walls of my own)?" (*Travelling in Amherst*, 23). Ever the intrepid uphill optimist, though, he notes in September 1935, "Every kind of life costs something. The price I pay is different from the price others pay because my life itself is different. I cannot escape payment, but I can make my life worth payment" (*Travelling* 44). Later, in his fifties, airy optimism turns to a kind of sparse sumptuousness and simple splendor, as he regales an anonymous guest with a feast of organic delights. His entry overflows like a menu that gives "rich evidence" of his garden's cornucopia: "potato salad flavored with my parsley and basil"; beets marinated in "tarragon vinegar from my tarragon"; "elderberry jelly . . . tinctured with some of my thyme"; "rhubarb shortcake . . . flavored with my spearmint"; and "Fredonia grapes from my vines were on the table at the close of the meal." Thus, he concludes, "a very poor man can play the almost-munificent host" (*Travelling in Amherst*, 76).

Eventually, his journals became a venue for personal declarations. During his period of most severe want (1952), he wrote of poverty as a "paradox," a "game and an art," a liability he tried to alleviate and an asset that, "in Thoreau's phrase," he tried to "cultivate" (*Travelling in Amherst*, 81). Most noticeably, his journal entries abound with figures of his annual income, tax totals, and other bits of bookkeeping. Alongside these account balances, his identity as an "independent and unpopular writer" emerges. "Though I live far below the American Standard of Living, I am not impoverished or pitiful," he declares, adding, "Few writers have more propitious conditions under which to write." Still, he writes, only "strict economy" gave him the ability to live "solvently and happily," a car-less carefree life of regular three-and-one-half-mile walks to town, no paid entertainments, and soybeans as his staff of life. "If I had more, I would give more," he observes. "If I earned more, I would have more. If the earning of more did not mean the losing of still more, I would earn more" (*Travelling in Amherst*, 86). In this ultimate turn of phrase, Francis reveals the full spectrum of his narrow existence, and the capacity of his artistic outlook to match the prodigal economy of nature. For Francis, poverty was both an inevitability and a choice, a fatalistic hindrance and his key to the kingdom.

In 1970, Michael Hamburger observed in *The Nation*, "By providing statistics of his income and expenditure, of the plants he grows and eats, of the things he buys and does not buy, Robert Francis shows how frugality can work in practice on the East Coast . . . and how its practice can lead to happiness, 'fulfillment and control'" ("151"). The year after Hamburger's statement, Francis published his autobiography, *The Trouble with Francis*, with the University of Massachusetts Press, in which he plotted his development as a writer for whom economy was an art.

In March 1939, Francis recalls, he was awarded $475 for the Shelley Memorial Award (the equivalent of over $7,400 in 2008), though he remembers that "he didn't know what to do with the money." Almost carelessly, he recounts how he lent most of the sum to local friends who needed it to pay for accidents and other bills, less out of generosity, he writes, "than a lack of imagination" (*Trouble with Francis*, 21). In the section that covers the post-Depression 1940s, he writes of being "torn between two economic and philosophic principles": to make "full use of everything [he] possessed and to get rid of everything [he] didn't need" (*Trouble with Francis*, 29). In chapter 8, "Lean Years for a Writer," he remembers a six-year period during which he didn't own a car and walked nearly four miles to Amherst with a "green cloth bag" in which he carried groceries and library books. Though he admits he yearned for a "cloak of invisibility" to take away the embarrassment he felt from being seen along the side of the road, he soon overcame it and found his walks to be a daily delight for mind and senses (*Trouble with Francis*, 85). Chapter 9, "Food," reviews his decision to abstain from fish, fowl, and flesh as an "ethical as well as an economic concern." Readers certainly grasp the extent of his economic straits while skimming the solo Thanksgiving dinner menus he enjoyed by himself over the years. For example, in 1948 "half of grapefruit" is listed as the entrée. In 1949, for dessert, he lists "homemade candied grapefruit peel" (*Trouble with Francis*, 100). His locally publicized "Spartan-Epicurean diet," as he calls it, was soon noised abroad and bred rumors, one in particular that he ate thistles. His fertile imagination, unencumbered by unnecessary calories, led him to compose a fictional letter to Thoreau (dated circa 1945–47) called "Soybeans for Walden," in which he lists the economic and nutritional praises of the soybean and jokes that Thoreau could have lived to be ninety on an altered diet and that perhaps Thoreau left Walden to "get some square meals" (*Trouble with Francis*, 103). And finally, in chapter 18, "Economy," Francis defines his habitual thrift as a "little common sense with a dash of the sportive," and he contrasts himself to Thoreau, who "had a much bigger story to tell . . . , since he had a much smaller budget to live on." Amid accounts of cutting his own hair, disconnecting his telephone, and scraping together guest lecture stipends and honorariums, he arrives at the point that he can say he learned "to cultivate the art of living without money." Thrift has "esthetic as well as economic

value," he finishes, noting that "thrifty" means both "frugal" and "fruit-
ful, two opposite connotations that derive from "the same Latin word"
(*Trouble with Francis*, 216, 217, 219).

"I keep confusing *ascetic* / and *aesthetic*," poet Lance Larsen writes
in "Questions for My Daemon." Though Francis never said it in such
express terms, what can be said now is that he discovered, partly
because of his life circumstances and partly by choice, not a confusion
between but a fusion of aesthetics and asceticism, or "aescetics." As a
practical artistic credo born out of necessity, the "aesceticism" Francis
practiced demonstrated how fine art can spring from hard times and
how the priceless riches of nature provide a bottomless wellspring of
material—both conceptual and physical—from which the penniless
ecopoet can draw enough water to sustain himself nearly into his nine-
ties. Clearly, Francis's surrounding native landscape and climate, as
well as his dependence on and embeddedness within the ecosystemic
processes in his bioregion, fashioned him into the writer he became.
This association exacted its cost but paid dividends. As David Young
writes, it was from "living in near-solitude most of his life . . . on a
miniscule income," and from "learning how to make do with little and
live off the land," that Francis "transferr[ed] the lessons of simplicity
and independence into adroit, tough-minded lyric poems and into lucid
prose" (David Young, "Robert Francis," 64).

As it turns out, while life deprived him of material comfort, that depri-
vation turned him toward the refulgent bounty of his immediate natu-
ral surroundings, in which he found a feast for both the body and soul.
The little that nature provided proved to be more than enough for the
ecopoet whose life and literature echoed the complex economics of sky,
river, field, tree, mammal, fish, bird, stone, and insect.

A PLACE FOR SPACE

In Francis's archival material, a unique sheet of paper reveals how
closely his literary subjects and writing processes were connected to
the environmental and cultural concepts of *place* and *space*.[2] This sur-
viving scrap, titled "Geographical locations of poems many of which
are in *Stand with Me Here*," aligns two columns: one for poem titles
and the other for the location at which the poem was written, or to
which it refers. This curious remnant lends new layers of meaning to
Francis's poetry, resonations not easily grasped in his *Collected Poems*.
On this sheet Francis associates many of the poems with interna-
tional geographical locations: "Hermit" and "Socrates" match up with
"Greece," and "Unanimity" sits opposite "Atlantic or Mediterranean."

Others show how specifically and intimately he wrote about more local places. Rather than just "Amherst," his poems "Bronze" and "The Stile" are linked to "Hop Brook, South Amherst" and "Hadley Road, Amherst." Francis provides no commentary on the assembly of his list, nor does he reveal its function. In the absence of an expressed purpose, however, this list demonstrates that Francis saw poetic creation as intimately connected to the faraway places he visited as well as the familiar locales he called home. In this rudimentary checklist, readers sense the global inseparability and intimacy that infused Francis's life and writing, a belief that poetry is as connected to the earth as are trees, burrows, ecosystems, and climate. In short, it shows Francis's poetry spanning the globe.

That Francis compiled his list as a conceptual or aesthetic standard of measure, or as an idle bit of writerly whimsy—rather than a historical or factual document—becomes apparent if we consider the publication of his poem "Shelley" in conjunction with his biographical and geographical wanderings. "Shelley," an economical encomium to the notable British Romantic, uses geocentric descriptors to depict Shelley in terms of the earth's processes, as if melting Shelley's literary persona directly into the ecosystemic source of inspiration that broadly defined British Romanticism: "Each had her claim. / To each he gave consent: / To water, his liquid name, / His burning body to flame, / to earth sediment / and the snatched heart, his fame / to the four winds. He went / As severally as he came—/ element to element. / Each had her claim" (1–10). In Francis's *Collected Poems* "Shelley" appears as having been published in his 1936 collection, *Stand with Me Here*. But in his autobiography, Francis records that it wasn't until his 1957 fellowship to Italy that he actually stood near Shelley's final resting place at the Cimitero Acatolico "to study the strangely divergent and contrasting graves" of Shelley and Keats and to go "more than once to the Keats-Shelley Memorial by the Spanish Steps" (*Trouble with Francis*, 112). Thus these questions arise: Did Francis write "Shelley" in the late 1950s and later choose to assign it an earlier slot in his *Collected Works* in 1976? Did he write it in 1936 and visit Italy later coincidentally—or as the result of a lifelong wish? Are any connections between his poem and travels incidental at most? While wide-ranging answers to these questions exist and might represent a fruitless and trivial line of inquiry, Francis's personal records, together with his place- and space-oriented poetry, affirm what Michael P. Branch has written, namely that "geographical space cannot be culturally neutral" (Branch, "V.E.C.T.O.R.L.O.S.S.," 4).

Many of Francis's poems operate on the energy generated by a place/ space dichotomy or dialectic, but a select few illustrate this more clearly than others.[3] In "Identity," an early poem in *Stand with Me Here*, the speaker stops abruptly in mid-stride on an alpine ramble, mystified by a footprint in the sand, only to wonder if the print is perhaps his:

> This human footprint stamped in the moist sand
> Where the mountain trail crosses the mountain brook
> Halts me as something hard to understand.
> I look at it with half-incredulous look.
>
> Can this step pointing up the other way
> Be one that I made here when I passed by?
> This step detached and old as yesterday—
> Can it be mine, my step? Can it be I?

In this terse lyric, the speaker experiences a half-determined meta-state, poised between identity and "unknowability." Francis's woodland hiking trail functions as the emblem of a wilderness place/space dialectic in that the uninhabited space of its sandy banks literally becomes stamped with the mark of a human footprint, thus momentarily slipping from space to place status, from unconnected to infused in the personal history of the walker. Referring to Yi-Fu Tuan, J. Scott Bryson observes, "If we think of space as that which allows movement, then place is pause; each pause in movement makes it possible for location to be transformed into place" (Bryson, *West Side,* 9). In "Identity," Francis's hiking speaker pauses literally in a woodland setting that previously represented outdoor freedom, only to discover, paradoxically, that the wilderness freedom he had supposedly gained is suddenly lost in the fact of his very presence. The speaker's incessant questions indicate that the identity of which he speaks—his own, the identity of nature, or both—has been lost in the seeking and, because of the compromising footprint, can now only be seen as a commingling of human and natural identities. "What begins as undifferentiated space becomes place as we get to know it better and endow it with value," Tuan observes. "Thus, enclosed and humanized space is place" (Bryson, *West Side,* 9). While Tuan does not indicate that this process of enclosure and humanization constitutes an overly positive or negative phenomenon, the speaker in Francis's "Identity" clearly remains unclear about his feelings toward the mark he has placed on nature and his inability to identify his identity in a wild space turned place. "As we move into space," Bryson indicates, "we discover, ironically, the extent of our limitations, rather than our freedom" (Bryson, *West Side,* 17).

Two poems from *The Sound I Listened For*, "Distance and Peace" and "True North," approach the place/space dialectic from the perspective of a human speaker who contemplates the relationality and relativity of supposedly fixed cosmic points of origin. These poems illustrate how, according to Tuan, "space is more abstract than place," and they address the twenty-first century observation that, according to Bryson, the concept of space appears to be "lesser-discussed" than place in the expanding canon of ecocritical literature (Bryson, *West Side*, 20). "Distance and Peace" meditates on the abstract quality of cosmic and natural spaces and shows how non-human spaces, while constantly in danger of being claimed and defined by human presence, remain aloof, distant, mysterious, uninhabited, and unreachable:

> Go far enough away from anything
> In time or space (and space is only time)
> And you have peace. The clashes of the stars
> Do not disturb the starlit night of earth.
> And earthly wars if they are old enough
> Make restful reading to a man in bed.
>
> And so with distance that is neither space
> Nor time. The grass we walk upon is peaceful.
> We can lie down on it and go to sleep,
> Being too far above it ever to feel
> The toil and competition of the roots,
> Their struggle, slow frustration, and defeat.

Here, Francis's speaker ruminates on the proximity of human and non-human wars, as well as moments of human and non-human peace and their relationships. The space of outer space, though understood from a human perspective to constitute a battle or clash, remains largely undefined and incomprehensible to earthlings. The wars fought and lost by grass roots, though occurring inches from the reclining slumberer's back, remain mostly in the realms of the abstract, due to their inaccessibility. Thus, in "Distance and Peace," while the human subject appears to contact the abstract spaces in the earth and cosmos, he fails to convert them into meaningful, human-defined places.

In "True North," the abstract spatial concept of the compass direction is exposed as a false notion, an ever-flitting will-o'-the-wisp that can never be defined as a human-endowed place. "Nothing that we can follow is true north," the speaker avers, "being a point in pure geometry" (1, 3). The speaker contrasts the concrete north on which humans can stand to the abstract north that serves only as a notion. "So I understand," he says, "true north / Does not stay true / but slowly travels in a circle too" and concludes that "it would be true to say true north / does not exist / Except to the extreme idealist" (4–5). To Bryson this clash, or "harmonization," between place and space produces the spark that marks the inception of poetry. "Poetry happens," he argues, "where place meets space" (Bryson, *West Side,* 21). The process of crafting ecopoetry will, Bryson says, "deemphasize the ego" and assist humans in "achieving place- and space-consciousness" and an "immersion of the self in the natural community, not by leaving the ego behind or by becoming a transparent eyeball, but by recognizing that we are members and citizens of a . . . 'land-community' to which we belong but will never master or fully comprehend" (Bryson, *West Side,* 22).

Finally, and perhaps most definitively, "A Broken View," the lead poem in Francis's 1936 debut collection, *Stand with Me Here,* captures the yin-yang relationship between space and place, an always-in-flux rapport that Tuan defines as "interdependent" and a "complex interplay" (Bryson, "West Side," 16). More philosophically and environmentally engaging than technically or linguistically captivating, "A Broken View" structurally signifies as well as conveys the speaker's message. The first weathered but mostly intact eleven-line stanza of slightly ragged iambic pentameter sets the outdoor scene. "Newcomers on the hill have cut the trees / That broke their view," the speaker says, initially neutral (1–2). Adopting a first-person plural perspective that recalls E. A. Robinson's tone of collective distance and communal commentary, the speaker reports, "An afternoon ago we stood with them / and saw their view," which includes "an unbroken sweep to south," "open places of pasture on the hills," and "sky among the clouds" (3). At the close of the opening stanza, the speaker proclaims the view "enough / For anyone to love for all a lifetime" (9–11).

If the first stanza represents the more systematic human presence and perspective, the chaos of the natural order erupts and dislodges the second stanza with tectonic tremors. While the meter struggles to remain constant, extended lines shoot eastward like overgrown weeds, and the eleventh line trickles out a single spondee: "In fall" (11). In the middle stanza, the speaker's environmentalist view supersedes the superimposition of egocentric logic, order, and human engineering imperatives in the face of the natural landscape. "Yet we were thinking . . . we loved a broken / View better, a view broken by trees," the speaker interjects (12, 14–15). According to the speaker, a so-called broken view of pasture and woods provides a superior experience because it is altered and framed in a variety of ways by the natural order of seasons and wild flora, rather than one cut to fit human needs and tastes: "A view that didn't give you everything / At once or anything too easily / One that changed as you went from window to window / and changed again as you went from month to month" (17–20). In allowing the second stanza to decompose organically, Francis signifies the transcendent moment of humanity's absorption into the more ecosystemically correct chaos of the world's natural vegetable and animal cycles of dissolution and reconstitution. Even snapping the phrase "broken view" in half like a dry twig across the end of a line communicates the poem's philosophy in micro: that broken views should be left broken.

In the final stanza, which suffers a central rift like a granite shelf sundered by a natural fault, the speaker emphasizes that his group of like-minded tree-lovers feels no regret and expels "no sighs" for the massacred trees, given their knowledge of the steady resurgence of forestation taking place slowly at the moment of the poem's utterance: "Young trees were growing / Among the stumps there even while we looked / Over their heads" (24, 27–29). Francis's persistent use of the first-person plural indicates that while the speaker remains somewhat complicit in these nearby acts of cosmetic deforestation, he invites his readers to share in a point of view that resists further thoughtless acts of needless land clearing.[4] In the concluding lines, which continue to spill out and overrun the banks of the poem's once strict meter and structure, the speaker imagines a future time when the regrowth of the woods will restore the more pleasing original broken view to those who return to that spot. "Perhaps we had a view / Through time, like a view trembling through leaves," the speaker finishes (30–31). In the restored forest's futuristic state, the newcomers' grandchild returns to the original location to "see as we had seen, loving both trees / And view, loving them more each for the other" (35–36). In "A Broken View," the phenomenon of marrying the landscape of the earth to the interlocking landscapes of mind and spirit achieves the state of "interanimation," as Bryson terms it. "It is this process of interanimation," Bryson writes, "of recognizing the interdependent nature of the relationship between people and the worlds they inhabit, that enables place-making" (Bryson, *West Side,* 11). Such works, Bryson concludes, constitute an ever-growing body of "inhabitation literature," ecopoems that "encourage us to discover and nurture a topophilic devotion to the places we inhabit" (Bryson, *West Side,* 12).

This ethos of topophilic devotion and place-making surfaces again and again in Francis's published and unpublished works. The ongoing and ever-developing presence of the place/space dialectic in his corpus suggests that it was almost as deeply rooted in him as his DNA. In the inner landscape of his imagination and the outer landscape of his experience, making place from pure space became a daily event, as regular as breathing. One journal entry (November 1933), recounts a moonlit walk to a local football field and tennis court. "As I went down, . . . terrace after terrace," he remembers, "I reminded myself in my mysterious solitude of some Aztec priest descending a sacrificial mound." Here, making meaningful places from abstract space is not bound to the metaphysical laws that govern space, time, and matter. "Something about the flat surface looking off to hills and valley, the bright moon, the sounds coming clear from a distance took me back to Beirut," he finishes, "where on so many nights I walked across the campus by moonlight and felt the buildings, the mountains, and the sea" (*Travelling in Amherst*, 32). His use of the verb "felt" casts the exterior world as something to be felt imminently rather than perceived from a distance. Further, his entry suggests that the activity of authorial place-making acts as a bridge or continuum that gathers all places into one, a place palimpsest or taxonomy of places that generates meaning for human subjects.

In 1980, the year after the city of Amherst dedicated the Mill River footbridge in Francis's honor, Mary Young of *The Valley Advocate* interviewed him. In this interview, he describes how, after moving to Amherst in 1926 with his parents because of his father's ministerial appointment, he took six decades to settle down and adopt the town that had "chosen" him as a young man. Perhaps subconsciously, but also significantly, he uses language that evokes the place/space dialectic and refers to his life journey not as a linear vector but as one of "finding a geographic center, spiraling in on a certain place in a certain town in a certain state" (Mary Young, 24). Even as he approached his eightieth birthday, the aged eco-poet remained engaged in the process of finding his center place amid a void of human, animal, and geographic space.

Three decades previous, in *Traveling in Concord*, he concludes his book with a flurry of cosmic queries. "Like day and night, like the four seasons, the old questions return," he writes. "Are we mechanisms or free agents? Are we animals or children of God? . . . What is my place in the great context? What plot of earth is for my roots, and what is my hour of fruition?" (*Traveling in Concord*, 186). While Francis understood, or at least expressed, that finding one's identity in the universe was a timeless occupation, it was not a "place-less" one. If any artist, philosopher, mystic, native, or priest can be said to have truly discovered his or her place in the vast space of the earth and beyond, Francis came as close to finding it as anyone.

7

The Experimental Environmental

THOUGH FRANCIS'S LIFESPAN STRADDLES THE periods traditionally associated with American modernism and postmodernism, it is neither accurate nor advantageous to see him solely as a writer of metrical nature lyrics, nor as a free-forming guardian of the environmental avant-garde. Those who read the Francis canon discover a palpable tension in the author's hybrid stylistic consciousness, a pull toward the supreme delicacy and technical mastery of closed forms, meters, and rhymes as well as a more chaotic, anti-establishment drift toward scattering words and phrases from his cupped hands, as if generating poetry were analogous to blowing dandelion spores to the wind. Throughout Francis's career, reviewers struggled to classify him as either stoic stylist or X-factor experimentalist. Writing for the *Christian Science Monitor* in 1971, Victor Howes dubbed Francis a "traditionalist who experiments" (Howes). A decade later, Robert Wallace critiqued Francis's "Excellence" in *Field* and observed, "In an era of the Avant-Avant Garde, Robert Francis, who can be passionate without being puffy, is a poet daringly Horatian. *Ars celare artem*. The art is to hide the art" (Wallace, 15). Wallace's 1980s depiction of Francis as an anachronistic traditionalist in the age of experimentation, complete with a scanned diagram of the hexametrics in "Excellence," speaks to the ongoing difficulty critics encountered when trying to nail down Francis's dominant mode.

The least problematic position to take at this point in history would be to say that Francis was both poetic imitator and innovator, while remaining a terrestrial caretaker. Although copious evidence remains to suggest he aped his contemporaries and mentors—especially in his earlier poems, those he called "relaxed traditional"—it is also important to note the ways in which Francis branched out, invented new forms, and questioned the notion of poetic form itself, as if his poetics were tuned to the evolutionary processes of the earth (*Pot Shots*, 131). As early as 1936, journal entries reveal his fascination with the experimental prose in Gertrude Stein's *Narration*, which he calls "delightful nonsense" (*Travelling in Amherst*, 46). Later, in 1945, he playfully spars with the world over poetic terminology: "Wade Van Dore showed me some experimental verses of his the other day. He asked me if I had done anything experimental. I told him that every poem was experimental" (*Travelling in Amherst*, 61). Notwithstanding the difficulties associated with definitions of "postmodern"—not to mention the sometimes perceived incompatibility between postmodernism and environmentalism—a substantial portion of Francis's poetry calls for just such a reading, one that combines the experimental and the environmental.[1] In all, four categories of experimental environmentalist verse emerge from Francis's published poems: "mono-rhyme," "word count," "fragmented surface," and "silent poetry."

Early on, Francis cultivated the mono-rhyme ecopoem and modified it repeatedly. Several more disguised versions of the mono-rhyme variety crop up in his subsequent collections. In *Frost: A Time to Talk*, Francis recalls with satisfaction how Frost, after critiquing several mono-rhyme compositions in manuscript form, poured out praise for "The Wall" and "It Is a Simple Thing to Die" (later titled "Simple Death" in *Stand with Me Here*): "*You know how to draw a fine wire edge over your gizzard. You've got to do that*, he [Frost] said. (These mono-rhyme poems which I liked and like so much had brought no enthusiastic response from anyone else)" (*Frost: A Time*, 58–59). To twenty-first century readers, the most postmodern aspect of a Francis mono-rhyme poem is that it subverts the egocentric, human-engineered polyphony of the various stanzas, rhymes, and meters in the traditional nature lyric with a droning, ecocentric cadence that more correctly echoes the more plodding but environmentally accurate ecosystemic biorhythms of insect wings, birdcalls, sunrises, stream currents, and wind through dead tree limbs. This verbal and aural innovation, while it tends to grind on the human ear used to lyric compositions of greater variety, succeeds in deprivileging the human presence and foregrounds a textual signification that mimics the earth's repetitive forces, elements, and sounds. While the traditionally formalistic nature lyric wrests nature from its organic state into an unnatural but more musical textual form, Francis's mono-rhyme poems let the text of nature drone naturally in a linguistic construction that more correctly represents the recurrent patterns in landscapes and outdoor phenomena.

In a sample of four Francis mono-rhyme ecopoems—"Monotone by a Cellar Hole," "The Wall," "Walls," and "Simple Death"—readers not only apprehend immediately the pulsing, heavy-limbed throb of words and end-rhymes, but they also sense an embedded strain of Robinson Jeffers's "Inhumanism" at work in Francis's sinewy, insistent lines. Not to be confused with "inhumane" or "anti-human," Jeffers's "inhumanist perspective," Max Oelschlager reminds us, "redirects attention from the human toward the inhuman and 'the beauty of things'" (Oelschlager, 246). "Monotone by a Cellar Hole," a slightly eroded seventeen-line arrangement of iambic pentameter, hammers away at the speaker's ultimate inability to comprehend life's vastness from his position near a human-made cellar hole, which is gradually being overgrown by native vegetation, thus erasing the local human presence from memory: "The sweetfern and the hardhack here have grown / More than all other living things have grown" (7–8). In addition to unapologetically ending successive lines with identical words, Francis inserts internal rhymes to heighten the impact: "The sun shines down on stone. It must have shone / No otherwise than this when it first shone. Wind blows as lonely wind has always blown, / a monotone not quite a monotone" (3–6). Writing in the mid-1930s, William Rose Benet groped toward describing the inhumanist presence in "Monotone for a Cellar Hole." "Behind the foreground definition," Benet observed, "there is an elemental background, felt if not necessarily sketched in, dominating such poems" (Benet, 30). Early on, this elemental background of which Benet speaks was elementary to Francis.

Everywhere in Francis's mono-rhyme poems, the insistent tone hums, and everywhere the language redirects the reader's gaze and ears to the "inhuman" beauty in the steady pulsations of the earth's tints, aromas, and echoes. "The Wall," a dehydrated tetrameter sonnet with an introductory nine-line stanza and a concluding quintet, passes for an inhumanist "Ozymandias" in that it chronicles how builders, "small / As walking flies," construct a wall that crumbles to nothing under the constant weathering of environmental forces: "Now only flies are there to crawl / Over the stones. There is no wall. / The highest stones were first to fall. / Stone after stone they fell till all / the builded stones had fallen—all" (10–14). "Walls," a free-standing sextet anthologized in Auden's *Faber Book of Modern American Verse* (1956), examines the dual nature of human-made barriers that prevent a "passerby" from beholding over-aggressive land development but also bar entry to natural landscapes: "Out-of-sight they say is out of mind / The walls are cruel and the walls are kind" (5–6). Alan Sullivan's response to "Walls," which he dubs "Manichean," typifies reactions from readers raised on the polychromatic polish of modernist lyrics: "What could be more brutal than these end-stopped lines," Sullivan writes, "each one grammatically walled from its neighbors?" (Sullivan, par. 15). "Simple Death," however, a far less brutal arrangement of thirteen faulty tetrameter lines, presents an ecocentric view of human demise, with human beings not at the center of the universe but somewhere near the periphery: "There need be no wide watchful sky / To watch one die, nor human eye. / No inauspicious bird need cry / At death, nor flying need it fly / Otherwise than birds fly / Whether or not some man will die" (4–10). Here, Francis uses an ecopoetic lever to move humanity from a position of dominance to a metaphysical meta-state where, once on equal footing with the biosphere, it inhabits the swirling matrix of the natural order.

In some cases, Francis's mono-rhyme compositions camouflage themselves with pararhyme and challenge the notion of the modernist "closed" text in favor of a more "open" postmodern textuality, while maintaining an inhumanist perspective. "Invitation," for instance, repeatedly urges curious acquaintances and friends not to postpone a hibernal visit to the speaker's woodland retreat: "You who have meant to come, come now / With strangeness on the morning snow / Before the early morning plow / Makes half the snowy strangeness go. . . . / You who were meant to come, come now / If you were meant to come, you'll know" (1–4, 9–10). Remarkably, when Francis, blind and in his eighty-fifth year, recited "Invitation" at the beginning of Henry Lyman's *Poems to a Listener* broadcast, he did so without a printed manuscript, from memory, fifty years after the poem was written and published. When Lyman asked him about his use of the word "strangeness" in the poem, Francis admitted he could have used other words then replied, "I don't know why I used 'strangeness.' Choose your own word" ("A Poet's Voice"). With the passage of fifty years, then, Francis's mono-rhyme "Invitation," infused with the redirecting influence of Jeffers's inhumanist perspective, serves as a prime example of the postmodern American ecopoem, in that the "text" of the poem transcends the fixed historical and spatial confines of the printed page to include the poet's memory, landscape, and seasonal changes, as well as the poet's open invitation for readers to modify its wording decades after he began its composition! "You who have meant to come, come now," the speaker chants. "When only *your* footprints will show, / Before one overburdened bough / Spills snow above the snow below" (5–8). In many ways, the poem's central figure serves as a commentary on the nature of experimental environmentalist verse—the undisturbed palimpsestic text of snow; its white-on-white layers of intertextuality both transparent and opaque; the solitary writer in the cabin waiting to read how the reader's footprints will rewrite the latest draft of winter's recurrent text; the overburdening and spilling of one author's text into another, ceaseless, without beginning or end. Somehow, within a monochromatic combination of modern style and postmodern content, Francis's mono-rhymes produce endless varieties of interpretations and open-ended meanings.

From mono-rhyme, I turn briefly to Francis's "word-count" poems. Though Francis claims in his autobiography to have invented the form outright, his word-count compositions bear a striking structural resemblance to the Imagist works of Pound and Williams. In his succinct autobiographical section devoted to explaining word-count, Francis self-consciously mocks his initial inner compulsion to both innovate and escape the copycat clutches of his contemporaries. "It amuses me to remember that when I hit upon this technique I was uneasy lest it be stolen from me," he recalls. "After a few word-count poems were in print under my name, I felt safer. But far from wanting to help themselves to my invention, my fellow-poets didn't even notice what I was doing" (*Trouble with*

Francis, 127). Though it is difficult to establish Francis as the true and absolute inventor of word-count, his *Poems New and Selected* (1965), which contained his largest selection of word-count poetry, predates similar mathematically structured poems by his contemporaries, such as May Swenson's "Four-word Lines" (1970) and Louis Zukofsky's *80 Flowers* (1978). Strictly speaking, Francis defines word-count as a means of "controlling the length of a line of poetry . . . by the number of whole words," which creates a "static effect good for certain subjects" (*Trouble with Francis,* 127). Fueled on playful postmodernist paradox, Francis no sooner explains his invention than he turns it on its ear: "If you ask what value there is in a fixed number of words in each line of a poem, I answer None" (*Trouble with Francis,* 127). Francis's tongue-in-cheek critique of his own invented form did not stop him, however, from authoring a steady stream of word-count poems, many of which pursue an enlightened worldview from the standpoint of non-human subjects—animal, vegetable, and mineral.

First, two standout environmentalist word-count poems, "Dolphin" and "Cats," draw the egocentric American reader from a comfortable position of seeing the world as an object to be possessed to a more ecocentric position of marveling at the motion, simple grandeur, and evolutionary miracle of animal and vegetable life. "Cats," a three-word-per-line poem that eerily mimics William Carlos Williams's "As the cat," describes the fastidious and stealthy locomotion of domestic cats but then, without warning, transports readers to a wilderness setting: "Wall-to-wall / Carpet, plush divan / Or picket fence / In deep jungle / Grass where we / Can't see them / Where we can't / Often follow follow / Cats walk neatly" (13–21). Visually, especially with its opening and closing lines being identical ("Cats walk neatly"), "Cats" resembles a more traditional lyric, but its verbal repetitions, swerving non sequiters, syntactical insouciance, and emphasis on the animal wilderness presence over human domesticity steers it toward postmodern environmentalism. As part of the printed body of the poem, Francis includes a footnote that reveals the source of the poem's first line: "Line 1 is quoted from Olaus Murie, Field Guide to Animal Tracks, The Peterson Field Guide Series (Boston, 1954), p. 113" (22–23). This textual trace, while perhaps innocent, constitutes a postmodernist moment of Kristevan intertextuality that demystifies the act of writing and achieves what Jennifer Ashton, in reference to Marjorie Perloff's "literalism," calls "laying bare the device," a technique of "using material forms . . . as an active compositional agent, impelling the reader to participate in the process of construction" (Ashton, 3).

The intricate "Dolphin," an almost mystical and numerological arrangement of seven septets of seven words per line (perhaps a forefather to Donald Hall's "The Ninth Inning" with its nine nine-stanza parts, nine-line stanzas, and nine-syllable lines), compares the mythic status of dolphins to "actual-factual" dolphins and their place in ecosystemic reality and human consciousness (4). In the closing stanzas, the speaker celebrates the dolphin's "large brain intricate as man's / And slightly larger," then closes, "Nothing less than forgiveness dolphins teach us / If we, miraculously, let ourselves be taught. / Enduring scientific torture no dolphin has yet / (With experimental electrodes hammered into its skull) / In righteous wrath turned on its tormentors" (36–37, 43–46). In the poem's finale, the phrase "righteous wrath," as an environmentalist concept, is similar to Terry Tempest Williams's description of Edward Abbey as "our sacred rage"—a holy anger, a violent but sacrosanct opposition to unholy acts of murderous environmental abuse ("Terrain.org Interviews"). To the average reader's eye, "Dolphin" appears to be little more than an American free-verse composition, though its practically invisible word-count technique is used effectively to make more visible the horrors of experimental science. In writing about Robert Frost's "Spring Pools," Glenn Adelson and John Elder have emphasized the concept of "invisibility," or "visibility/invisibility tension," a notion they use to describe ways in which poetic structures mimic, or parallel, geographical structures and characteristics. In the case of Frost's "Spring Pools," a *douzain* (or sonnet with its concluding couplet "buried"), Adelson and Elder compare the poem's more invisible components to the "invisibility" of rising and lowering water tables that "bury" pools of water in summer but reveal them in spring (Adelson and Elder, 9–10). Similarly, Francis's "Dolphin" employs an invisible poetic technique to expose the socially and culturally invisible practice of animal torture that masquerades as scientific progress.

Two additional word-count poems, "Delicate the Toad" and "Stellaria," employ what Helena Feder has called the "rhetorical mode" of environmental apostrophe, or the Romantic tendency to directly address nature and natural phenomena (Feder, 43). "Delicate the Toad," a three-word-per-line composition, pictures a toad that "sits and sips / The evening air," completely "satisfied / With dust, with / Color of dust" (2–6). Having sketched the contemplative solitude of the toad, the speaker turns to the birds overhead, directly addressing them: "Laugh, you birds / At one so / Far from flying / But have you / Caught, among small / Stars, his flute?" (10–15). "More than merely a poetic device," Feder argues, environmental apostrophe is "nothing less than a practice of awareness that emphasizes the process of human perception, the interconnectedness of all things. . . . It is a practice intended to call our attention to the poet's interaction . . . with the natural world and the understanding that interconnectedness is in itself a form of reciprocity" (Feder, 44). To address nature, Francis demonstrates, is to address one's self.

Likewise, "Stellaria," a five-word-per-line poem, finds the speaker addressing the wildflower whose "frost-white / Petals and plum-purple stamens" project a colorful beauty invisible to everyday onlookers (1–2). "Who / But the botanist ever sees?" the speaker questions, then points out, "Your foliage is weed familiar / But your flower is almost / Like a fairy princess invisible" (4–5, 6–8). With a Keatsian about-face, however, the speaker's lament turns to laudation: "Better so. Those who escape / man's notice escape man's scorn" (8–10). Though nearly invisible, Francis's apostrophic mode of address invites readers to see the natural world's intricacies and reciprocities. "Inherent in any invocation of the natural world is a recognition that reciprocity is embedded in the very interconnectedness of all things, in an awareness of the sensitivity and multiplicity of those intricate connections," Feder observes (Feder, 44). Francis's arresting apostrophic addresses, while outdated, some might argue, redirect readers to the immediacy of the natural world and its sometimes invisible changes. "Apostrophe demands that we, as readers, allow ourselves to be enveloped in an ever-present moment," Feder points out, "a space that . . . emphasizes the poet's linguistic and physical connection with the subject of the poem" (Feder 54).

Though others may have ignored Francis's word-count, he made up for their lack of commentary by writing and speaking about it himself. In the case of "Icicles," he constructed in *Pot Shots* a ludic postmodern critique of the poem in the form of a mock neoclassical-Platonic dialogue between "Damon" and "Daphnis." The essay, "Word-Count," finds Daphnis asking, "Seen any of the so-called word-count poems Robert Francis writes?" After hearing Daphnis describe the technique, Damon replies, "I couldn't get excited about that" (*Pot Shots,* 20). The middle of the essay traces the split-hair quibbling that Damon and Daphnis exchange as they argue over word-count's putative use or uselessness, comparing it to the counting of rhythmic feet and the syllabic verse of Marianne Moore. Then, having copied "Icicles" from a library book, Daphnis quotes the entire poem, calling it "typical Francis":

Only a fierce
Coupling begets them
Fire and freezing

Only from violent
Yet gentle parents
Their baroque beauty

Under the sun
Their life passes
But wait awhile

Under the moon
They are finished
Works of art

Poems in print
Yet pity them
Only by wasting

Away they grow
And their death
Is pure violence. (*Pot Shots*, 21)

Every paradoxical element in "Icicles" combines to cast it as an experimental environmentalist poem—the lack of syntax and punctuation, followed inexplicably by a single icicle-point period at the end of an icicle-shaped stalactite of stanzas; the ephemeral aesthetics; the claim that icicles, as art, are finished only in the moment that nature "begins" to "end" them in the process of melting; and the juxtaposition of beauty as violence, and art as erosion and eradication. In the *Pot Shots* interview with Tetreault and Sewalk-Karcher, Francis considered his word-count poems bridges that allow the poet to remain in touch with the world. "For me a poem is something made as well as something said," he revealed. "My poems confront the actual, recognizable world that we share with one another. However imaginative and original my vision and interpretation of that world, I do not want to lose connection with it" (*Pot Shots*, 131–32). Word-count became one of many ways in which Francis, through poetry, reinvented and renewed his relationship with the singular and multiplicitous earth.

Francis's third ecopoetic innovation, "fragmented surface," employs a linguistic and syntactical liberation to more accurately match physical poetic features to the earth's topography. Francis's 1974 book, *Like Ghosts of Eagles*, overflows with fragmented surface poetry. Seeking a freer range of expression and, as he writes in his autobiography, "groping for greater emotional impact," he wandered away from the "connected discourse" of his early more traditional verse and began to let words and phrases dabble on the page like wind and rain (*Trouble with Francis*, 128). True to his instincts, Francis chose geographical and elemental subjects for his fragmented surface ecopoems. In "D," he plays on the fragmentary materiality of his title's alveolar plosive signifier in order to describe the physical features of cliffs that overlook the sea:

Deliberate
Determined
Undeterred

> The cliffs stand out against the sea
> Cliff beyond cliff, each cliff a D.
>
> The sea assaults them as the sea will;
> The cliffs still stand there, stand there still
>
> Dauntless
> Defiant
> Undismayed. (1–10)

Here, the physical shape and sound of a single consonant communicates the dissonance and consonance created by the concussions and collisions of shore and surf.

In some instances, Francis's fragmented-surface poetry mimics the broken surface of rivers and streams. "Like Ghosts of Eagles," a poem arranged on one level, as Donald Hall observes, to "celebrate language," also uses its visual appeal to emulate the shape of rivers, their rocky eddies, countercurrents, rivulets, sidestreams, and meandering banks (Hall, "Two Poets," 121). In a way, the punctuation-free text resembles an entire watershed:

> The Indians have mostly gone
> but not before they named the rivers
> the rivers flow on
> and the names of the rivers flow with them
>   Susquehanna   Shenandoah
>
> The rivers are now polluted plundered
> but not the names of the rivers
> cool and inviolate as ever
> pure as on the morning of creation
>   Tennessee   Tombigbee
>
> If the rivers themselves should ever perish
> I think the names will somehow somewhere hover
> like ghosts of eagles
> those mighty whisperers
>   Missouri   Mississippi

In addition to mapping the shape of a river system on the page, Francis separates himself from the twentieth century by hearkening back to a time when indigenous populations in pollution-free environments sought not to own, but only to name, the land they inhabited. Futuristically, and somewhat hopefully—though realistically—this lyric transports readers to an as-of-yet-unexperienced epoch in world history when, even in the presence of polluted and dried-up riverbeds, the memory of America's native cultures and their terrestrial stewardship will not be erased.

Similarly, in "The Mountain," whose title avalanches into its first line, Francis layers embedded words and phrases together like rock strata to illustrate the degree to which human culture and geophysical features are inextricably connected, in reality as well as in the imagination:

> does not move the mountain is not moved
> it rises yet in rising rests and there
> are moments when its unimaginable weight
> is weightless as a cloud it does not come
> to me nor do I need to go to it I only
> need that it should be should loom always
> the mountain is and I am I and now a cloud
> like a white butterfly above a flower.

This pressurized text represents the complex interconnectedness of human and non-human biospheres, and it calls into question overly simplistic views of ecology in the company of overly complex views of human psychology and imagination. Glen A. Love, in "Revaluing Nature," has written of this simplicity/complexity dichotomy. Love observes, "It is one of the great mistaken ideas of anthropocentric thinking, and thus one of the cosmic ironies, that society is complex while nature is simple. . . . That literature in which nature plays a significant role is, by definition, irrelevant and inconsequential. That nature is dull and uninteresting while society is sophisticated and interesting" (Love, 230). In the simply titled "The Mountain," Francis constructs a detailed, fragmented poem that demonstrates, through linguistic artistry and free-play, the faulted strata and complex workings of the earth.

In other fragmented-surface ecopoems, such as "Blue Cornucopia" and "Blueberries," Francis questions an overly simplistic humanity's inability to perceive the innumerable, complex layers in naturally occurring colors, the color blue being one of his most often treated subjects. "Blue Cornucopia"—a shattered sonnet—demonstrates that the simplistic, anthropocentric concept of "blue" as a singular color fades when held against nature's fantastically complex color wheel on which countless prismatic shades and degrees of blue shimmer, radiate, and waver:

> Pick any blue sky-blue cerulean azure
> cornflower periwinkle blue-eyed grass
> blue bowl bluebell pick lapis lazuli
> blue pool blue girl blue Chinese vase
> or pink-blue chicory alias ragged sailor
> or sapphire bluebottle fly indigo bunting
> blue dragonfly or devil's darning needle

blue-green   turquoise   peacock   blue spruce
blue verging on violet   the fringed gentian
gray-blue   blue bonfire smoke   autumnal
haze   blue hill blueberry   distance
and darker blue   storm-blue   blue goose
ink   ocean   ultramarine   pick winter
blue snow-shadows   ice   the blue star Vega.

Using the same technique, "Blueberries" dissociates readers from tradi-
tional syntactical patterns in order to foreground the concrete and com-
plex nature of its colorful subject: "faint blue dark blue their bloom /
untarnished wild rose ferns / the hayscented brushing a granite boulder /
blueberries highbush lowbush / and silence old-fashioned silence / a single
towee 'drink your teee'" (1–6). Invited to stray from normal lexical pat-
terns, readers hop from image to image, bird-like, seeing the landscape
in the chaotic, disjointed way it would be viewed firsthand. Debauchee of
blue, the speaker chants, "white clouds sky pasture blueberries / a bird's
booty / a boy's occupation / stain of blueberry on a boy's mouth / and all
around the blueberry-colored hills / dark blue faint blue blue / beyond
blue the farther the fainter" (7–12). Dissatisfied with simplistic egocen-
tric notions of color, Francis fills his bucket, overturns it, and stains the
page with the infinite degrees of nature's indelible, multitudinous hues.

In addition to writing about the complexity of nature's colors, its end-
less rivers, and its chiseled cliffs and outcroppings, Francis used frag-
mented surface to combat destructive trends in over-urbanization, as
well as the decline and pending eradication of certain insect, animal,
and bird species. As early as 1950, in his collection *The Face against
the Glass*, Francis was feeling his way toward a poetics that challenged
the clash of human engineering and delicate ecosystems. "Superior Van-
tage," for example, places the speaker on a bridge over some swimmers
in "innocently-azure water." The speaker wonders, "From the superior
vantage of our knowledge / (Knowing the pollution of the river) / We ask,
crossing the bridge, the ironic question: / How clean can one be washed
in tainted water?" (3–6). With a similar spirit, the fragmented-surface
poem "City," which hinges on some darkly comical punning, is shaped
like a line of skyscrapers tilted on its side so that the horizontal cityscape
becomes a tree that was bulldozed to make way for a freeway:

In the scare
city
no scarcity
of fear
of fire
no scarcity

of goons
of guns
in the scare
city
the scar
city

In a way, this de-evolutionary nuance in structure directs a prophetic voice of warning from the late twentieth century to the present moment, now that land-grabbing overdevelopment and overpopulation have become urgent issues in need of redress.

In the fragmented, asyntactical "Suspension" (whose shape recalls William Carlos Williams's cantilevered stanza arrangements), the poem's unstable structure literally collapses in the moment of reading. Today, a poem such as "Suspension" appears uncannily prophetic, especially when read in the context of the appearance of honeybee colony-collapse disorder, which has recently become a pressing reality and a pending sign of global ecocatastrophe. The elliptical form of "Suspension" eliminates the human presence completely and leaves readers with a scene of haphazard pollination, unpolluted by humanity, perhaps in some future post-human setting:

Where bees bowing from flower to flower
In their deliberation
Pause

And then resume—wherever bees
Cruising from goldenrod
To rose

Prolong the noon the afternoon
Fanning with wings of spun
Bronze

Sweetness on the unruffled air
Calore and *colore*
Where bees

Not satisfied with egocentric views of nature as simple and humans as complex, Francis snapped sentences across his knees like kindling, peeled the bark from each word, and crumbled the fragments on the page as if casting lots with the gods and goddesses of nature. In order to engender healing and change, Francis used his fragmented surface poetry to penetrate the surface relationships of a broken society on a broken earth.

And finally: silent poetry. In my preface I referred to Francis's "Silent Poem" as the *ne plus ultra* of experimental environmentalist lyrics. While there may be more apt—and perhaps less hyberbolic—phraseology to describe "Silent Poem," as well as more deserving poems to bear the title, I feel that in many ways (his long narrative *Valhalla* notwithstanding) "Silent Poem" serves as Francis's crowning poetic achievement, his *minim opus*. "Silent Poem" was one of the first Francis poems I read, and so in my mind it conveys the quintessential Francis outlook and aesthetic: country solitude, meditation, stillness, the absence of human bustle, the slow progress and regress of time, or, more accurately, the absence of time itself in the presence of the earth's trees, rodents, weather patterns, and intermittent flourishes of sun and soil.

Synchronically, "Silent Poem" appears in *Like Ghosts of Eagles* (1974), the last singular book of poetry (not counting the *Collected Poems* and posthumous *Late Fire, Late Snow*) to be published in Francis's lifetime, twelve years before he died. Historically, it sits as a kind of stone marker, or trail's end—the poem he waited his whole life to write until he found it, resting there for him, in the dusk of his days. In various critiques, reviews, and other sundry publications, "Silent Poem" surfaces more frequently than any other piece as an example of Francis's essential aesthetic and subject matter.[2] Those who feature it appear both fascinated and baffled by it.[3] In most cases, critics and reviewers simply reprint the entire poem and, as if helpless to do much more, add little or no commentary, perhaps silenced by the poem whose words drive readers into a state of wordless aporia:

backroad leafmold stonewall chipmunk
underbrush grapevine woodchuck shadblow

woodsmoke cowbarn honeysuckle woodpile
sawhorse bucksaw outhouse wellsweep

backdoor flagstone bulkhead buttermilk
candlestick ragrug firedog brownbread

hilltop outcrop cowbell buttercup
whetstone thunderstorm pitchfork steeplebush

gristmill millstone cornmeal waterwheel
watercress buckwheat firefly jewelweed

gravestone groundpine windbreak bedrock
weathercock snowfall starlight cockcrow

Staggering in its simplicity, arresting in its arrangement, impossible to ignore—"Silent Poem" produces as much silence in readers as it draws from nature to enact its composition. That direct and hypnotic transfer of silence from nature to reader may indeed be just the point. Robert Shaw calls "Silent Poem" a work of "daring minimalism and amazing purity," an "unadorned inventory" that "evokes an entire landscape . . . and the life led in it" (Shaw, 86). In seeking to understand "Silent Poem," Donald Hall perhaps comes closest in labeling it a work of art that takes readers "deep into pure naming." Hall writes, "Actually among the most backward-looking of poems, . . . it takes poetry back to that condition of quiet and wonder when sounds were first formed and linked to objects of the world, when every word was fresh and a poem" (Hall, "Two Poets," 123). As Hall indicates, something primeval imbues "Silent Poem," as if the ecopoet in 1970s rural Massachusetts has become a neo-Adamic citizen on the first day in his new Eden, and the gathering and naming have begun anew.

One way of approaching "Silent Poem" is to see it as evidence that Francis was engaged in creating, or discovering, an entirely new genre of poetry. His published and unpublished writings suggest that silent poetry was an organic, intuitive concept he nurtured and developed for years. What, we might ask, could be more postmodern and experimental—and perhaps absurd—than silent poetry? An invention of that sort might seem as useful as leafless trees, waterless rivers, or peak-less mountains. As it turns out, Francis conceived of and wrote silent poetry well in advance of various ecocritics who later mulled over the value of silence, the literary arts, and the environment. In "Nature and Silence," Christopher Manes points out that in "animistic cultures, those that see the natural world as inspirited, . . . plants and even 'inert' entities such as stones and rivers are perceived as being articulate and at times intelligible subjects" (Manes, 15). In terms of the cultural paradigm Manes describes, Francis's thunderstorm, groundpine, and snowfall speak in a way that may seem inarticulate, or silent, to human ears tuned solely to human frequencies. "Our particular idiom," Manes continues, "has created an immense realm of silences, a world of 'not saids' called nature. . . . We must contemplate not only learning a new ethics, but a new language free from the directionalities of humanism, a language that incorporates a decentered, postmodern, post-humanist perspective" (Manes, 17). By stripping away the intrusive writerly "I" from the poem and by relying solely upon concrete words to speak silently, Francis accomplishes in art what Manes proposes in culture and society—"taking the silence of nature itself . . . as a cue for recovering a language appropriate to an environmental ethics" (Manes, 17).

At the same time, critics of experimental environmental poetry, such as "Silent Poem," would be quick to question the efficacy of such an innovation. After all, they might contend, what is it good for? How much can a pastiche of words accomplish? In reality, "Silent Poem" contains no *things*, no actual woodchucks, chipmunks, or jewelweed, but words that represent those things, a poetic aesthetic reminiscent of that practiced by William Carlos Williams and other Imagists. On this level of understanding, Plato's classic distrust of poets in *The Republic*—accusing them of artistic fraud since in writing poetry they produce thrice-removed linguistic representations of things rather than the true things themselves—would tend to undermine Francis's, and any ecopoet's, environmentalist project. David Gilcrest has addressed this issue, one that has become "increasingly familiar . . . to contemporary writers and readers of environmental poetry," that is, the need to understand the social, cultural, and environmental impact of "[t]he distinction between *res* and *verba*, between the things of this earth and our words for them" (Gilcrest, 18). Applying this distinction to Francis's "Silent Poem," we could argue that while Francis's list of concrete words does construct a representative barrier between humans and the actual natural world, on another level it raises the reader's awareness of the power of representations (and misrepresentations!) of nature, for good or ill, and, due to its postmodern ethos, encourages readers to privilege actual experience with the natural world over poetic arrangements. On a more visual, literal, and textual level, its stagger-step couplets represent the footprints of the silent walker whose meditative stroll conveys him through an outdoor corridor of arresting objects, sounds, and aromas.

It is likely that in writing "Silent Poem," Francis sought a dual objective: 1) to encourage the human need for unmediated experience with the natural world; and 2) to acknowledge the weakness of language, its shortcomings and ambiguities, as a means of living closer to nature. As an experimental environmentalist poem, "Silent Poem" calls for people and poetry to fall silent and behold the environment's experiment firsthand.[4]

Though it is difficult to tell Francis's tongue-in-cheek satire from his straight-forward sobriety, he dwelt repeatedly on the subject of silent poetry in his *Pot Shots at Poetry* collection, which provides additional background on his foray into experimental environmentalist poetics. "On the Exquisite Air" complains about the superfluous talk that accompanies poetry and poetry readings. "Why poets do so much talking in public I really don't know," the speaker declares. "Sometimes the commentary and confession come in such an engulfing stream that the poems are quite submerged and only dart in and out of sight like swimming fish." As a point of contrast, the speaker asks if a painter, after displaying his works, posts himself at the door to give "every visitor a personally conducted tour." Preferring the absence of cluttering commentary, the speaker concludes, "I should like poems hung, one at a time, like Japanese pictures, on the exquisite air, each poem surrounded by space and silence" (*Pot Shots*, 11).

In another microessay, "Energy," an intensely focused speaker asks, "Do words have secret energies waiting to be released like the energies within the atom?" By taking this angle, Francis's speaker appears to suspect, if not strongly believe, that the ultimate power and force of poetry pre-exists the human presence and that when a poet "writes" a poem, he is not making something as much as he is standing by and beholding its fruition. *In extremis*, the speaker suggests that poets can no more make poems than they can make acorns or oak trees. "If they [words] could escape their immemorial bondage to grammar and syntax and sentence structure and conventional human meaning and do for once what they really wanted to do," the speaker continues, "if they could make love to one another as free agents, would they bombard the reader with undreamed-of power?" Here, silent poetry is described as a return to purity, as a means of seeing language in its primal organic state, sans the cultural filter of humanity. "I am haunted by the vision," the speaker finishes, "equally unable to believe it true or to believe it untrue" (*Pot Shots* 156). Silent poetry, in Francis's conceptual world, is silent precisely because it requires no human voice to give it life. "Energy" depicts the life force of poetry as analogous to atomic energy—always already here, there, and everywhere—a potent, invisible force that becomes lethal only when human voices and motives begin to interfere.

It is interesting to note that Francis began to develop his experimental concept of silent poetry only a matter of years before William Rueckert published his "Literature and Ecology: An Experiment in Ecocriticism," the essay that purportedly first coined the term *ecocriticism*. In it, Rueckert's academic call "to develop an ecological poetics" mirrors Francis's homegrown breed of silent poetry, and some of Rueckert's ideas sound as if he borrowed them from Francis's "Energy" (Rueckert's 107). "A poem is stored energy," Rueckert declared, "a formal turbulence, a living thing, a swirl in the flow. Poems are part of the energy pathways which sustain life. Poems are a verbal equivalent of fossil fuel . . . , but they are a renewable source of energy" (Rueckert, 108). Here, Rueckert follows Francis, in that both writers presuppose the inaudible, humming power of poetry as a naturally occurring element, something that pre-dates the arrival of the human "voice" that gives it its audible quality. "Properly understood, poems can be studied as models for energy flow, community building, and ecosystems," Rueckert continues. "The First Law of Ecology—that everything is connected to everything else—applies to poems as well as nature" (108, 110). Given the historical proximity of these two writers, it would not be too much of an impropriety to suggest that we should include Francis's silenced genre of silent poetry as a contributing factor in the eventual rise of ecological literature and studies—that his silence actually gave a voice to ecocriticism in its infancy.

In the short essay "Silent Poetry," Francis reveals the full scope of his struggle. In writing about "his creation," he appears as baffled as the next bystander, like a farmer perplexed by the sprouting corn he planted in his own field. This distanced regard of his own concept suggests that Francis truly considered silent poetry to be something that occurred organically in nature and that as a practicing ecopoet he had simply stumbled upon its presence, like a hiker happening on an apple tree in the forest. "The idea of silent poetry or silence in poetry used to puzzle as well as fascinate me," he begins. "I wanted such poetry to exist but I couldn't quite see how if by 'silent' was meant 'wordless' or 'nonspeaking.'" Though his tone is elusive (is he serious?), his word choice, elliptical claims, and lapses into the passive voice are revealing. Why did he want silent poetry to exist? Who—or what—"meant" what was said? Reading this opening paragraph, the satirically minded observer might be tempted to wonder if the satirical rogue had spent one too many nights alone in Fort Juniper. Still, the directness and the deliberation with which Francis treats his radical, experimental genre suggests that he was in earnest, that the silent poet-less poem was something he felt could help humanity find a more harmonious relationship with its surrounding ecosystems and environments—if we would just lower our voices, stop shouting, listen, and look around.

Ironically, the speaker remains vigorously vocal in his pursuit of silence. "How is silence achieved?" he presses. "A very short poem, like a cry on a still night, makes the surrounding stillness more vivid by breaking it. . . . A poem that presents an object or scene or situation without comment approaches the silence of painting or sculpture." Ultimately, the musing microessayist asks the question that could silence all arguments concerning silent poetry: "Why couldn't a silent poem . . . mean one that impresses us with what it leaves unsaid?" (*Pot Shots*, 60). Throughout his quest, Francis remained playfully serious concerning silent poetry, the genre that sought to eliminate the human presence completely from the act of writing nature poetry. And though few poets scorned the critics more than Francis, throughout his later years, many writers and reviewers picked up on his insistence that poetry could generate itself, like weeds and wildflowers, without the presence of human beings.[5] The conspicuous absence of the egocentric pronoun "I" amid all the leafmold, honeysuckles, hilltops, and buttercups in "Silent Poem" demonstrates this drive toward generating a pure poetic art untainted by humanity. As such, Francis's twentieth-century breed of author-less poem aligns itself with what Lawrence Buell, in reference to the work of Dominic Head, sees as a necessary "deprivileging of the human subject" in the ongoing twenty-first century dialogue between the "broader Green movement" and postmodern theory (Buell, *Future*, 10). Silent poetry, then, brought Francis as close as he could get to using the imperfect and cumbersome medium of language alone to convey to others the raw essence of the simple wonder, majesty, awe, and power in the world around him.

It is, of course, with obvious irony that in the twenty-first century we are forced to lament—and wonder at, but not necessarily accept—that the author of "Silent Poem" should be silenced for so long, due to a life of writing and subsisting in wooded solitude on the rural fringes of American literary life. This is not to say, at a period of historical environmental crisis when we need now more than ever to hear what the earth is saying, that Francis, though silenced by time, needs to remain silent forever.

8
Valhalla

AN EXAMINATION OF FRANCIS'S CONTRIBUTION to American ecopoetics must certainly include a reading of his long narrative poem, *Valhalla*. Though it might seem backwards to conclude a study on Francis with a poem he wrote at the beginning of his career, the poem's main theme—environmental apocalyptics—centers on the meaning and significance of endings, both cultural and ecological. Thus, this remarkable narrative poem serves, paradoxically, as the perfect temporary terminus in a discussion of how Francis's work may begin to shape the future of humanity and the earth.

Curiously, Francis provides little information about why he chose to dedicate so much sweat to *Valhalla*; likewise, he remains relatively taciturn in his writings and conversations about its completion, as if his shorter lyrics meant more to him and defined his writer's ethos more accurately. "Today I finished my long narrative poem, *Valhalla*," he notes simply on July 28, 1937, "on which I have been working for nearly three years." Then, with a trace of regret, he adds, "Whatever becomes of the poem, I am sure that I am a better educated man, and perhaps a wiser one, than when I started it. This poem was written in the Hopkins' house, at the Jones Library, at Mrs. Boyntons's, at Mrs. Newkirk's, at my parents', and lastly, here in Cushman" (*Travelling in Amherst*, 51–52). To carry on with *Valhalla* so long, and to carry it to so many temporary locations, suggests that Francis had internalized a kind of authorial demon that wouldn't let him rest until he had purged it in the act of writing. As records show, he must have exorcized the narrative succubus because, as was the case with his novel writing, he never made the attempt again.

The compulsion to author the great American narrative poem infected Francis early on. Hints that he would have to scratch the three-year itch arise in conversations with Frost, and in Francis's private notations. "Frost distrusts long poems," he writes in 1935, and a footnote added in the 1970s reveals that in 1933 he had already taken a crack at a decidedly environmentalist strain of the narrative form with *Johnny Appleseed*, an epic on the "half historical half legendary Jonathan Chapman, a poem never finished" (*Travelling in Amherst*, 42; *Frost: A Time*, 52). It is also possible that as an author reading and writing in the wake of a long-standing tradition of American narrative poetry—with such notable predecessors as Crane's *The Bridge* and Robinson's *Cavender's House*—Francis

166

felt compelled to compose his own in order to join the ranks of his peers, an American poet's rite of passage, so to speak. Robert Shaw has compared *Valhalla* to the longer poems of E.A. Robinson and Jeffers, the "short novels in verse . . . with a strong regional interest" from the 1920s and 1930s (Shaw, Robinson, 77). Virtually unknown and unread today, *Valhalla* remains an important example of the American and international tradition of "environmental apocalypticism," as Lawrence Buell has called it—narrative works that have re-shaped and continue to redirect the outlook of world cultures whose toxified and quickly vanishing natural resources presage sudden and disastrous ends for human and non-humankind alike (Buell, *Environmental Imagination*, 285).[1]

Early and late in the twentieth century, *Valhalla* garnered hyperbolic praise and ho-hum soft-pedaling from various readers.[2] Francis himself preferred to refer to the collection as a whole, rather than the single poem, and he recalls that on its publication it "made no stir anywhere," though it did produce a cherished letter from Frost (reproduced on a full page in Francis's autobiography), as well as the Shelley Memorial Award for $475 in 1939 (*Trouble with Francis*, 21). Critical quibbling aside, the six-part *Valhalla*, if not in recognition of Francis's sheer feat alone, deserves to be reread and more fully appreciated for its environmentalist thematics, apocalyptic message, and artistry.

Part I introduces readers to the characters, tone, and story that involve a Vermont family's struggle to exist on its ancestral farm in the face of environmental forces and climate change, as well as to the occupational sociocultural hazards of the modern world. At the opening, Leif, the son of John and Edith, celebrates his twelfth birthday, for which he is given a present: an ax. Armed and eager, having grinned and run his "thumb over the blade," Leif, accompanied by his older and younger sisters, Eden and Johanna, chops down the "old dragon," an apple tree at the periphery of their farm, under the watchful vigil of their parents (1.18). After Leif dispatches the old dragon apple tree, John, his father, in the company of Sylvester, the gray and weathered grandfather, coerces Leif into beheading a hen as a test of Leif's manhood, although Leif's "unacknowledged fear of blood"—a task Leif completes despite the "spurting blood" on his clothes makes him nearly faint (1.36, 97). Around an aromatic feast of roasted chicken, golden pumpkin pudding, and Eden's fresh bread "dark with butternuts and hickories / Seasoned a year from old Valhalla trees," the family raises a convivial and seasonal toast to the life and legacy of "Leif the Dragon-Killer" (1.134–35, 150). The evening ends with Eden and Leif pulling on the chicken's wishbone. Traumatized and visibly altered—"His eyes were shadow in the candlelight"—Leif, enervated by a spirit of futility, tries "to make his own prong break," but still comes away with the larger wishbone portion in his fingers (1.213).

Before blowing out his candles, he stuns his family with his reply after they ask him what he wished for: "I didn't wish" (1.199).

In terms of environmentalist and apocalyptic thematics, Part I uses symbolic language and consistently recognizable tropes to depict Leif as a divided figure torn between devotion to his father and his sensitivity toward the non-human world. In the opening stanza, the narrator describes Leif, whose Norse name rotates on a homophonic pun for "leaf," as the offspring of both the machine and the garden: "the hickory wood / Matched his hair . . . / the grayness of the steel / Resembled something gray in his blue eyes" (1.6–9). A hybrid human figure of both the earth and the industrial age, Leif slays the "old dragon" of the dead apple tree (a nominative conflation of both mythic power and natural savagery), but struggles to defeat the internal dragon of his psyche, which bends to the will of his father's patriarchal dominance, his father's mesmeric eyes "reaching down in his / With a power to draw out what was hidden there" (1.74–75).

In Part I, ecofeminists might react critically to John's heavy-handed tutelage that steers Leif toward becoming a domineering, land-raping protégé of his father. In the doorway, as Leif's parents watch him hack at the old apple tree, Edith's eyes grow dark, "as water darkens under a sudden cloud," and John looks on approvingly at the "so familiar certainties" in his son's imitative actions; quickly, Leif adopts the role of the patriarchal land-destroyer: "They saw him point to the place / Where the girls must stand—Stand there—and the girls obey" (1.33–34). In his discussion of ecofeminism and the "Magna Mater," Max Oelschlager describes the divided consciousness between ancient and modern outlooks that characterizes the relationships between male and female figures, as well as human and non-human entities, an observation that applies to *Valhalla*—the split between the "Paleolithic consciousness that saw the earth as feminine, contrasted to Cartesian-Newtonian mindsets that saw the land as a feminine resource to be raped, seduced, used, and discarded—whore-like" (Oelschlager, 310). Likewise, Terry Gifford, in his work on the genre of the "post-pastoral," constructs a triadic relationship that links male figures to female figures and the land they both inhabit, an "issue that ecofeminists in particular have helped us understand better—that our exploitation of our environment has emerged from the same mind-set as our exploitation of each other" (Gifford, 84). Though marked by boyhood innocence in Part I, Leif begins to trade innocence for an inheritance of domination and exploitation over his feminine, animal, and environmental "others" as he becomes hypnotized by violence.

Part I includes two additional significant plot elements: the withering of Leif's imagination, and environmental forces as potent characters in the action. Midway through Part I, following Leif's execution of apple tree and hen, the narrator describes a scene in which Sylvester's wife, Ruth, the grandmother, gathers clean laundry that "danced and dazzled on the line," when Leif appears (1.114). "Leif saw only the clothes, / The empty, flapping clothes. Within the hour / Something had happened to his make-believe" (1.115–117). Like Coleridge's mariner, Francis's Leif experiences an inner death, the death of his imagination, due to his crimes against nature.[3] This sundering of the human imagination and the environment, or Cartesian "hyperseparation," as Greg Garrard has referred to it, characterizes the supposed dominance of Euro-American culture to the extent that "humans are not only *distinguished* from nature, but *opposed* to it in ways that make humans radically alienated from and superior to it. This polarization . . . often involves a denial of the real relationship of the superior term to the inferior" (Garrard, 25). Thus hyperseparated from nature—the mariner's crossbow now his shiny ax—Leif no longer feels the need to dream or wish for anything, even with the biggest part of the wishbone in his hand: "But Mother, why should I have to make-believe? / . . . I wanted an ax and now I have an ax" (1.212–14). Also like the mariner, Leif is required to bear physically the symbol of his crime by eating Eden's bread that has been seasoned with the almond-like "insides of apple seeds" from the murdered tree (1.179–80). The savory and symbolic ingredient mystifies the entire dinner party (including Eden, who baked the bread). Thus, the text turns a moment of environmental sacrifice into an act of social sacrament, a gesture toward establishing accountability towards Leif's actions against the land.

At key moments in Part I, environmental forces take on the role of actors and actresses in the drama. Immediately following the execution of the apple tree and hen (and the death of Leif's imagination), the narrator describes an apocalyptic rumbling outdoors: "The high wind blew all morning over the hill / And afternoon, blowing loose leaves from boughs, / then almost blowing them back. A few birds / blew from tree to tree, and one big crow blew off the hill, / using his wings only to steer. The sun / went down in wind though not a second / Sooner for it. A few things wind can't reach" (1.123–29). As in *The Rime*, nature spins in a turmoil, the elements seethe, and the world wars with humanity. After John raises a toast of "long health" to their "hill and home," a similar atmospheric disturbance nearly blows the door down: "As if the wind had been waiting for this pause, / It rushed at the windows and door in a fiercer effort / To force itself inside" (1.155–57). Because of Leif's actions and John's inculcation of patriarchal, anti-environmentalist habits in his son, nature raises a warning howl both peremptory and prophetic, a token of tempests to come.

Part II opens with Leif's sudden attack of appendicitis in October and ends with his convalescence at home two weeks before Christmas. This medical emergency, the plot element and its details, derive from Francis's own battle with appendicitis while he was a student at Harvard, including the specific treatment of feeding the patient warm water from a teaspoon, which Leif receives fictionally and Francis received in real life (2.335–36; *Trouble with Francis*, 178). In the opening tableau, Edith watches Leif as he chops the felled apple tree into firewood, "busy with his ax and dragon / Although it was no dragon to him now" (2.229–230). As if stricken with a curse, Leif slumps to the ground, and John and Sylvester rush him via horse and carriage to a country doctor's home. Soon Sylvester returns with news of Leif's appendicitis, and Edith and the girls wait at home.

Three days later, John returns with news of Leif's recovery, but instead of working alongside his wife and daughters to ready the home for Leif's return, he turns strangely moody and incommunicative: "There were kitchen consultations every day. / Without consulting, John took down his shotgun, / Cleaned and oiled it" (2.364–66). Though the speaker's words do not specify it, John's actions depict him as a reactionary who takes revenge on nature as a result of nature's attack on his son's life. Soon, John returns with his kill, a sight that horrifies his youngest daughter, Johanna: "Slung from shoulder to shoulder and from the pole / Hanging by feet bound by a rope a deer. / The limp ears drooped, the head dragged on the ground, / And blood was on the throat. Johanna saw / And ran from seeing and hid in the dark hall closet / And closed the door" (2.377–83). Unexpected, however, is the presence of a "dark" seventeen-year-old stranger named Judd, who has killed the deer, a doe, whose "dark eyes stared as if they knew, / They understood but could not move to answer" (2.385–87).

Judd's ethnicity remains ambiguous; he is described only as "dark," and the death of his father appears to have left him hardened and bitter. The family invites Judd to dinner, and symbolically Judd occupies Leif's place at the table while he discourses on his work in the printing trade. With Leif in his room, Johanna shyly steals in and returns Leif's ax to him, after which, in another room, the doctor, Doctor Moor, lectures John on the dangers of raising his family in such primitive, remote conditions. "Are you not leaning on your luck / A little too securely, a little too / Serenely?" Doctor Moor challenges (2.538–540). In a stream of criticisms that label John a fundamentalist hermit, Doctor Moor accuses John of "running away" from civilization, of misanthropy, and of denying his children the chance to "fall in love": "Yet people are not so dangerous after all / As none" (2.560–61). "You, your family are the cream of the earth," the doctor says, "yet we cannot live on cream" (2.553–55). Unmoved, John reveals his deterministic view: "One can hide from danger and think that he is safe, / Or one can live a life that's worth the danger / That's bound to be" (2.547–49). Part II ends with Leif and his family enjoying a venison supper as snow falls outside and Christmas approaches. Johanna delivers to her brother the last innocent, but perhaps racist, line: "The dark boy sat in your chair at supper, Leif" (2.610).

In Part II, interspersed among the human drama, Francis continues to increase the level of environmental description and action, so that the natural world becomes an even more potent actor, or player, in the narrative. This sustained degree of "biotic egalitarianism," as Buell labels it, operates in *Valhalla* in a variety of ways, some stylistic, some more substantive (Buell, *Environmental Imagination*, 303). Inexplicably, Edith undergoes a supernatural transformation, and she has her eyes opened to a vision of nature bleeding and weeping as a result of her son's tree killing and her husband's complicity in slaying the female deer. As she runs from the house to help Leif, who has fallen in the act of chopping wood, a flood of blood bathes the earth and then vanishes: "The maple leaves were blood on tree and ground / And blood on Leif she saw before she saw / There was no blood, but only Leif on the ground" (2.248–50). Literally, in Edith's eyes, her "bloody Leif" has joined the "bloody leaves." After John and Sylvester rattle away with Leif to the doctor, Edith struggles but fails to explain to Eden and Johanna her vision of the wounded earth and ends up speaking to the environment instead: "The window,—I was standing there, she said / Less to the girls than to the evergreens / Or to the clouds over the evergreens" (2.279–80). With Leif gone, Edith succumbs again to the vision of the bleeding earth and nearly faints: "She stood there seeing / Nothing beyond a familiar bit of wood / And sky—until she was aware of blood / On trees and on the ground, and of herself / About to fall" (2.310–12). One sign of *Valhalla*'s drift toward biotic egalitarianism is the way that select characters, under the influence of a higher power, see the pending death of the non-human world through human eyes.

In Part II, to a stronger degree than in Part I, the biotic egalitarianism that undergirds Francis's apocalyptic vision generates vivid extended passages that describe the tumultuous and subtle dramas of nature. At these moments, nature's struggles and lamentations move from backdrop to foreground, demanding greater attention from the reader and putting themselves on equal ground with the human action, perhaps even "upstaging" or minimizing the human action. At times, these passages achieve a startling dramatic effect, a kind of environmental dramatic irony—obvious to readers, but invisible to characters in the drama. At other times they border on sentimental anthropomorphism that can strike twenty-first-century readers as trite. Following Edith's near-fainting episode, the narrator counterbalances the human inner drama with the story of the outer, non-human drama: "Night was early / moving among dark trees, / Trailing their shadows down across the farm / Till all was shadow. A night that could bring rain" (2.314–16). Not only is night capable of an almost animal-like locomotion, but also a primal release of emotion in the form of rain-tears: "Reluctantly / And late, the scattered drops at first, then pause / More drops, and then the delicate noise of rain / On leaves on trees, and roofs, and leaves on ground, / Bringing another day" (2.350–354). More than bland atmospheric setting or backdrop, these environmental details function as players in the drama.

In terms of plot and characterization, Part II reveals important details about Leif's family and their cultural outlook and socioeconomic status, more specifically, their sense of time, place, and history. When Leif travels to the hospital, Sylvester and John take him in a "carriage" and keep the "horse to a steady trot / Along the wheeltracks" (2.260, 271). Such a picture would tend to suggest that Leif's family inhabits a nineteenth-century rural farm, and yet, when Johanna waits for Leif to return, she watches "car after car crawl into sight / Along the ribbon of road" that approaches their farm (2.477–78). Two possibilities exist then: 1) Leif's family refuses to join the modern world; or 2) they are too poor to afford modern conveniences and modes of transportation. Doctor Moor's observation that John was lucky to find both himself and the "surgeon in Vermont" at home when they were needed suggests that John is still living in an era of traveling country doctors, whereas the modern world has moved toward physicians that keep regular hours and require patients to attend office visits and planned surgeries (2.530). Francis's jarring clash of time frames advances the reader's feeling that the apocalyptic end for Leif's family and the world they know is still approaching. In other words, the more John's nineteenth-century view of life and landownership spars for supremacy over the modern rise of technological advances and commercial values, the more convinced readers become that one way of life must soon override the other and that somehow the delicate balance of the surrounding ecosystem will either flourish or perish, based on the outcome. At the close of *Valhalla*'s Part II, the "old ways" and the "new ways" overlap one another and reflect one another's differences against the unchanging earth. At this point, readers sense that an evil influence is eroding both the earth and the lives of Leif's family members, and that a slowly building battle for supremacy between two authorities—John's and nature's—will result in the sudden emergence of a new epoch for both the land and its inhabitants.

The spring fling that invokes Part III, following the somber fall and winter motifs in Parts I and II, signals a sudden upswing in mood, setting, and imagery. Amid "pasture bluets," "rue anemones," "windflowers," and the "first green grass," Leif and his family, in a pre-harvest bucolic outing, dance and frolic with Audhumbla, the family milk cow, to the tune of John's "running hornpipe" music in a vernal display of ecstatic dance and song that evokes the purity, classicism, and simplicity of a Virgilian eclogue (3.613–15, 621). Following this opening return-to-nature sequence, the narrator then gives a detailed account of Eden's devotion to growing sweetpeas, followed by a fast-forward jump to Doctor Moor's unannounced visit to the family's hayfields in mid-July. As a token gesture of reconciliation—and so he can covertly examine Leif's scar from the surgery—Moor briefly and awkwardly assists the men in loading hay on their hayrick and accompanies them to their swimming hole, though, unlike Leif and John, who unabashedly strip and plunge in nude, the doctor removes only his shirt and dashes "handfuls of water on face and arms" (3.753). Back at home, Doctor Moor, "playing the connoisseur," feasts on an organically grown pea supper with Leif's family, after which the narrative leaps ahead to August and an erotically and ecologically charged scene that centers on Eden, alone, nude, in the family apple orchard: "Heavy with heat / Herself she slipped her light dress off, and sighed, / And touched a fallen apple with her toe / Tentatively, then poked it out of her way, / And sighed again and lay down in the grass / Under a tree" (3.838, 880–85). Roused by a pending thunderstorm, Eden dresses and races home, where she, Leif, and Johanna strip naked, join hands, and "run in the rain" that deluges the farm unabated, while Ruth and the parents watch (3.993). Then suddenly, and perhaps melodramatically, at the decrescendo of the rain-swept bacchanal of innocence and ecstasy, Sylvester hobbles up and reports that lightning has struck and killed Audhumbla where she stood and sought shelter under an elm.

In Part III, readers begin to see that what *Valhalla* lacks in driving action, complex narrative, and compelling structure, it compensates for with its subtle suggestiveness, multi-layered environmental imagery, and evocative description. In Part III in particular, Francis's extended philosophical and image-driven passages, which at times imitate Homeric epic catalogues, embody another one of the "four modes of perception" that characterize the environmental apocalyptic narrative—that of "conflation," according to Buell (Buell, *Environmental Imagination*, 305). The most pervasive and noticeable conflation Francis employs connects land and sea imagery. In the climactic rain-bathing scene, Francis situates Leif, Johanna, and Eden on water and earth simultaneously: "The three began to race . . . / Out over the pasture-sea. They splashed the puddles / To surf, and the grass was seaweed under their feet" (3.1004–05). Watching them, Ruth, to herself, calls the children "blessed fishes," having "no word for anything half-fish, / Half-human" (3.1007–09). As Sylvester watches the thunderstorm lash the Vermont landscape, he says to

John, "A storm at sea is another thing from a storm / On land. Are your mountains rock or are they water?" (3.941–42). Earlier, when Doctor Moor strides into the hayfields to assist the men, he sees the hayrick "dip and rock / Over the rough ground like an old galleon, / Sylvester navigating from the bridge, / While over his head and over the hills great clouds / Were a surf along the wide beach of the sky" (3.737–41). When John invites the doctor for a "dip," John refers to the secluded woodland pool as their "inland sea," or, as he puts it, "The nearest Valhalla ever came to the Ocean" (3.746, 751). Then, at the swimming hole, when Doctor Moor gazes with somewhat homoerotic fascination at the naked bodies of father and son—the "generic human body, genus homo," constituting in his clinician's mind the "ingenuity of nature"—he, like Edith earlier, has his eyes opened to a surreal vision of nature: "He almost saw, had felt the threat and was saved / From seeing something obscene, misshapen, monstrous, / Something fished from the mile-deep midnight ocean—a pale and mutilated octopus" (3.774–75, 777–80).

Though Alan Sullivan misreads this octopus imagery as a Freudian instance when the phallic and "private obsessions of the author swim murkily into view," this admittedly disjointed conflation of land/sea imagery is shocking to a certain degree (Sullivan, par. 14). As an apocalyptic literary device, it suggests that the ultimate end toward which Francis's narrative is driving will not be one of violent cataclysms and explosive annihilations, but one of ultimate reunification. Literally, through *Valhalla*'s veiled allegory, sea and land are reuniting, returning to their state of pre-Christian, pre-human one-ness. With new eyes, Francis's human characters glimpse this movement of the earthly elements toward a state of originary unity when land and sea were one.

The apocalyptic mode of conflation in Part III operates on many levels. Its complex interweavings include conflations that, among many others, can be described as human/non-human, earthly/heavenly, and mythical/biblical. At the pea supper, when Johanna and Eden tell Doctor Moor that they prefer to eat raw peas "right out of the pod," Moor replies, "That makes you rabbits, / Doesn't it? Doesn't that make you rabbits or deer?" (3.854–55). However, such a child-like observation produces no laughter in the children, who receive his inquiry as "a question / Touched with fear" (3.864–65). The children's reactions suggest that, in their world, such human-animal crossovers are part of the life they live and nothing to be treated lightly. Before the supper, a grave skepticism informs Doctor Moor's questions to John and Leif in the hayfields. "You plan to spend all of your time in living?" the Doctor challenges John. John replies, "We calculate there will be time enough / For other things when we are dead." "Why die at all?" the doctor chides, calling John's family "inhabitants of heaven." He follows, "Where is heaven if this is somewhere / Else?" Unnerved by the doctor's insistence, John says, "Leif hasn't come to heaven yet / In his geography . . . / He has studied only solid earth thus far, / Particularly this hill, this part of Vermont" (3.810–22). Here, in this dialogic sparring, Francis conflates traditional notions of heaven and earth and adds to his apocalyptic vision the question of why, historically, cultures have thought it necessary to divide a whole entity into two concepts.

Valhalla's apocalyptic vision forecasts a reunification to replace such false cultural divisions. In terms of the mythical/biblical conflations, examples abound throughout Francis's narrative. Eden's nude slumber in the apple orchard recalls Eve in biblical Eden, so intentionally, in fact, that Francis sends a black "shiver in the grass," or snake, to arouse Eden just as the "dragon-headed" thunderstorm passes over head, "flashing its forked fire tongue" (3.904, 907–12). Among such biblical allusions, Francis intersperses mythical elements from Norse sources (Leif, Valhalla, Audhumbla, Johanna); Greek sources (the "Damoclean danger" of apples over Eden's head, John's "Atlas-like" strength); and those of Germanic and Scandinavian origin (the "Nixies," or aquatic sub-deities, that Ruth has never heard of) (3.706, 881, 1007). By conflating biblical and mythical allusions, Francis draws readers from the present moment of historical and cultural schismatics to a time when nature was the only source of worship, before human cultures began to see the necessity of separating fantasy from doctrine, and paganism from Christianity. *Valhalla's* multiplicity of apocalyptic conflations invites readers to conceive of an era when the universe once unified land and sea, heaven and earth, and people and animals—and, by extension, of a future era of reunification.[4]

Part IV begins in May, turns to August, and then culminates in an autumn departure. In terms of action, the narrator now depicts Judd as a young man with a "shop," "printer's sign," "business card," and "roll of dollar bills in his pocket book" (4.1101–02). Among other things, Judd's occupation as a printer situates him as a character whose passion for advances in human communication technologies drives a wedge of increasing insensitivity between himself, others, and the natural world.[5] While Francis, as a writer, would have seen the art of printing and publishing as a necessity, Judd's characterization as the desensitized printing entrepreneur stems from experiences in Francis's life when he seems to have weighed heavily the impact of printed language on the environment.[6]

Though Francis only hints at Judd's erotic interests and development, it seems clear that Judd has been hiding himself in the woods and watching Eden's nude sunbathing from afar. Francis also hints at how Judd's mother discovers evidence of his nocturnal emissions: "His pillow mornings when she made his bed / Was proof of dreams, but the pillow never told her / What the dreams were. . . . What he did remember / After these months was the sting, the giddiness, / How he had had to shut his eyes for strength / To see again, and the sudden taste in his mouth" (4.1109–17). His voyeuristic views of Eden are mingled with the outdoors: "Naked, nothing hidden or withheld / From him, yet all oblivious of him, / Color of sunlight, blossom, fruit—the girl / Under the apple trees and looking up. / And he rigid within the edge of woods, / His hand grasping a tree, a heaviness / Like death upon him" (4.1118–23). Meanwhile, Leif and John erect a tower of "four spruces" that allows the family to read and stargaze and find seclusion (4.1169). From this observation tower, Leif happens to spy Eden and Judd as they rendezvous in the woods; Leif, "having never fought before," fights with Judd, is beaten badly, and returns to consciousness to see someone "fanning him," though the mysterious benefactor, perhaps a supernatural sub-deity or numen of the woods, vanishes once Leif awakes (4.1258). At Judd's printing shop, John confronts Judd and delivers the news of Eden's pregnancy; he tells Judd to visit him the following day at his home with a "plan" (4.1348). At John's residence, Judd says he loves Eden and will marry her, to which John replies, "A man may love the woman that he steals from" (4.1377). Unprovoked, John exposes his desire for Eden to remain on the "hill where she has lived since she was five," to which Eden, having entered the room, responds, "Why is this the only place to live?" (4.1395, 1422). At the close of Part IV, Eden and Judd depart for the village, and Johanna gives Judd a "little frightened kiss" (4.1468). Francis modifies the biblical expulsion scene, and instead of casting out Eve from Eden he casts out Eden entirely, dramatizing a banishment of nature at the hands of a man for whom printing, language, and commercialism have become the new religion, a substitute for engaging and preserving the wild sanctity of the natural world.

By the time readers reach Part IV, *Valhalla*'s non-human characters are vying for center stage. For example, this section opens not with human affairs, but with the landscape viewing the landscape. "The valley sees the pasture on the hill. / Below the pasture and above are woods / Up to the wooded peak up to the sky. / The valley sees the darkness of evergreens / Waiting above the pasture to come down / As other evergreens have come or wait / To come to darken pasture on other hills" (4.1057–63). The extraordinary—and perhaps disorienting—thing about this opening passage is that Francis describes topographical alterations from the perspective of the land that experiences those changes. In reality, barring cataclysmic events such as avalanches and mudslides, hills and evergreens and pastures take generations to change in the way Francis describes, according to human perspectives. But by writing about the changes in landscape as if they happen more visibly, Francis appears to condense time, magnify the scope of the landscape's transformations, and present a timeless, more cosmic view of the earth's cyclical lifespan— from the view of the earth and the universe, not people.

In addition to conflation and biotic egalitarianism, Buell has identified "magnification of scale and collapse of distance" as two of environmental apocalypticism's "major motifs" (Buell, *Environmental Imagination,* 305). For Leif and his family, the tower they build represents a last-ditch effort to transcend their collapsing world, an evanescent agrarian age and sheltered lifestyle that they cannot inhabit forever. A prosthetic (but not synthetic) extension that promises to elevate them above the hill that no longer affords a sufficient view, this tower of geo-Babel appears to magnify their views and collapse distances. The tower assists them in seeing stars, overgrown farms, and evergreens, but it has its drawbacks: "But seeing more and farther was less to them / Than seeing their own Valhalla from above, / The hawk's way. . . . / They saw their farm as something outside themselves. . . . / Progress or retrogress—/ They saw time moving over the hill from age / To age, from summer to fall to winter to spring, / From day to night" (4.1175–79, 1183–85). Rather than affording them a greater sense of personal stability and a grander scope of the surrounding earth, their human-made tower produces, instead, a greater degree of egocentrism and temporal instability. When Leif and Judd battle in the woods, Francis's condensed language enacts a magnification of scale and collapses distances: "Leif stopped. No question spoken / The one word, Snake! Then fists. The trees closed in. / Two worlds. One would be saved, the other lost" (4.1254–56). As a microdrama that accompanies and contributes to the approaching apocalyptic finale, the boys' fight is rendered in highly magnified dimensions, on the level of "worlds." Similarly, for Johanna, the unpredictable magnification powers of the tower transport her to a time and space warp when she ascends it solo one afternoon: "Johanna felt the hill slip back an age, / Or was it forward? And she the only one / Who saw and was. No farm was there, no man, / No woman, child. She was the last or first" (4.1323–26). The indeterminacy in the narrator's voice indicates that Johanna, for the moment, loses her existentialist grip on human-determined time and space. Through buried, the paired lexical signifiers "saw and was," as mirrored inversions, represent the degree to which human referents and meanings are being unraveled as readers near the culmination of Francis's apocalyptic narrative. Leif's family dissolves because it sees what was, not what will be.

In Part V, a Yeatsian gyre overtakes the narrative, its language, and its structure and propels it outward and onward, like a disintegrating blizzard of starlings or vortex of dusty leaves, into an ever-widening expanse that more accurately matches the chaos of the cosmos. For example, the passage of the seasons begins to accelerate. "Not mist, not rain, not even a rainy night / Could shut the hill in like the fall of snow," the narrator begins, indicating a winter setting. Then, the action leaps forward to "one day in June" at a time when a sixteen-year-old Johanna is tending her vegetable garden, followed by a warp-speed trip ahead ten months to the next April when the family learns of Leif's death at sea and Johanna anticipates Edith's death from the onset of what appears to be pneumonia (5.1469–70, 1584, 1730). As the narrative approaches its inevitable anti-resolution, Francis employs at random a mixed jumble of all four modes, or motifs, which coincide to produce a work of environmental apocalypticism: interrelatedness, biotic egalitarianism, condensation, and magnification.

In Part V, the linear progression of anthropocentric time shatters. The opening stanzas describe an elderly Johanna who enters a barn and is now "whiter than" the "old white cow" inside named Audhumbla—not "Audhumbla the first but only the second. / No, not the second, the third, at least the third" (5.1483–89). As the elderly Johanna talks to Audhumbla's descendant, a flashback vision of her girlhood skiing trips with Leif flares into her mind, then we leap forward to the spring when "Eden's boy was born" and an eighteen-year-old Leif, while chopping wood, asks, with John's blessing, to leave the farm because the hill has become "too cramped a place to live" (5.1530, 1548). The seasonal calendar continues to spin to "the summertime," with letters arriving from Leif and Eden— Leif has swum in the ocean and secured employment on a "sailing ship" in the Caribbean; Eden's "Little Judd" has sprouted teeth (5.1622, 1647). Mysteriously, Edith makes a secret and silent journey to the spruce watchtower, climbs it, looks around, and descends, then later requests that she knit Johanna a dress as "blue as the sea" for the time "when Leif comes back" (5.1696, 1705). Leif, however, never returns. Before the family receives a letter that reports his death at sea, Edith—again, the occasional visionary—runs from the house to embrace the image of her grown son, who appears to be "sauntering down the pasture toward the house," but who, unbeknownst to Edith, has passed away and sent his ghost home in his place (5.1732). In a trance, Edith travels to the spruce watchtower beneath "whiteness of mist and a moon low in the sky," catches a chill, and becomes bedridden under the care of an aging Doctor Moor (5.1806). "We die from what we love," the sixteen-year old Johanna tells Doctor Moor, as they sit together and endure Edith's final hours (5.1851).

Part V, in keeping with preceding motifs and themes, draws upon the opposing connotations in two words: interdependence and independence. Buell uses the figure of the web to illustrate this dimension of the apocalyptic narrative. "Just as the metaphor of the web of interdependence is central to the ethical force of the contemporary ecocentric critique of anthropocentrism," he argues, "so is the meta-phor of apocalypse central to ecocentrism's projection of the future of civilization that refuses to transform itself according to the doctrine of the web" (Buell, *Environmental Imagination*, 285). In Part V of *Valhalla*, readers begin to discern how certain characters have accepted the doctrine of nature's web of interdependence, whereas others have rejected it and sought through their independent rationale and energy to master nature via the stubborn application of a kind of environmental "will to power."

Ecofeminists would likely point out that those characters in *Valhalla* who lean toward interdependent habits and thoughts are mostly female; those that exhibit independent tendencies are mostly male. In depicting Johanna's garden, the narrator appears to take the side of the female characters, as if criticizing John's supervisory insistence that Johanna's garden conform to human strictures. In June, John stands looking "at the impeccable pattern of greens, / Serious-minded, sober vegetable greens / In geometric rows—lettuce, spinach, / Beans, carrots, cabbage, even potatoes, / And not one weed and not one flower either. / The leaves were lovely as the leaves of flowers / But loveliness was not their reason here" (5.1584–90). A militant, patriarchal perfectionism has been foisted on Johanna's garden. The narrator chides, "There was more to see / Than vegetables in strict unerring rows / If one had wit to read between the rows" (5.1593–95). Though Johanna, the sole child to remain on the farm, works better "than some boys do" (according to Sylvester), it is clear that to John she "couldn't take Leif's place" (5.1601, 1612). Though the narrator never states it explicitly, he hints that the onset of the coming apocalypse is somehow a cosmic result of John's unbending sexism and speciesism.

Contrariwise, Edith, the visionary, on her trip to the spruce watchtower the night she contracts a killer chill, sees the sprawling forests and fields in terms of their oceanic properties from her vantage point, an epiphany of global interconnectedness: "She was above the mist and looking down / And off at green dark islands washed by a sea / Of foam suspended, smoother than any sea" (5.1806–7). Similarly, Johanna, at the moment of Leif's death, finds that the crisis opens a portal of perception through which she sees her life and Leif's death connected to the geosphere and all living things: "Death was darkness and the beautiful deer. / Death was lightning, storm, the sea Leif loved / Drawing him down and down with open eyes" (5.1767–70). Between these variant gender polarities, Francis's narrator plants descriptions of nature's uncontrollable, inter-dependent transitions and movements: "The overlapping intermingling season / Of winter-spring when anyone is free / To call the weather either, neither, or both / . . . Birds coming back, snow going, but more to come.

/ Coldness and warmth clinging to each other / Like wrestlers wrestling
for the fun of it" (5.1522–1528).

The interlocking struggle of the wrestler provides a central image for
the way Francis feels that human and non-human lifeforms should inte-
grate themselves: in a symbiotic contest that produces positive growth
and development for both. The subtlety and strength of Francis's char-
acterization lies in his depiction of a farmer who has, ironically, labored
his entire life in *intimate outward* contact with the earth, and yet who
has not been *moved enough inwardly* to live in such a way that his habits
and practices reflect true intimacy and interdependence. Having taken
the wilderness as his enemy—as an unruly force in need of an overpow-
ering master—Francis's stubborn patriarch, John, is unable to triumph
in the wrestle with himself. Since he has sought to conform the organic
world to his will, rather than finding his place in the web of intercon-
nectedness, he has become entangled in it. As he kicks and flails to free
himself from a perceived entrapment in the web of the wilderness, each
tremor that travels down the tenuous silken threads that connect him to
the world's organic unities threatens to drop him, mummify him in his
violent thrashings, or else summon the loss of his family, his livelihood,
and the land upon which he supposedly relies.

In Part VI, Francis's human story reaches its tragic and fatalistic end
while the environmental and wilderness narrative that has paralleled it
for over two thousand lines delivers a decisive "non-ending"—a symbolic
continuation of action and organic drama.[7] As in previous sections, the
rapid and indistinct rotation of seasons accompanies the ultimate dissolu-
tion of John's family. The section begins in May with Johanna, now older
but of indeterminate age, visiting an arrangement of boulders that has
been overgrown with weeds and "small gray birches" (6.1879). The nar-
rator's tone, and Johanna's behavior, indicate that she is psychologically
unbalanced since she sits among the stone ruins of a farm settlement that
predates her family's settlement and talks to Leif and her mother, Edith,
as if they were still alive: "Do you remember, Leif? Don't you remem-
ber?—/ Mother, if you were I, if you were I—" (6.1892–93). Also, Fran-
cis describes the way egocentric time frames have dissolved and been
absorbed into the more cyclical rush of ecocentric non-time: "Gradual
growth, gradual decay, / And little else to tell the years apart. / Years
had once had shapes of their own like hills / Along the sky, But now they
blurred, they merged" (6.1883–86).

Valhalla's end is fatalistic, naturalistic, and tragic in human terms. It is
indifferent, and even triumphant, in non-human terms. On a "November
afternoon" under a "sky of steel," Johanna and John, both older, visit the
"old-stone almost prehistoric place" that had been a farm, where John
asks Johanna why she has not yet left the family homestead. Johanna's

insistence that Leif and Edith are still alive brings a reproof from John—"Girl, your mother is dead. Your brother is dead"—to which Johanna replies with an explosion of repressed questions about their lives and the land, questions John answers only with a bowed head and silence: "Why did you ever come to live on this hill? / . . . Why have I loved it if I was not to live here / Always, always? Why are you my father / If I am not to love you, if I must leave you?" (6.1925–30). Later, between "the fall of leaves and the fall of snow / The interregnum, age of bronze and silver," after John doesn't return from a wood-chopping expedition, Johanna finds him dead in the forest due to unstipulated causes (6.1973–74). After John's burial, "Aunt Johanna" receives a surprise visit from "young Judd," Eden's son, who stays with her for a few days (6.2040). Near the end, Johanna finds her father's black diary and, even after reading it, still finds herself unresolved about her family's history and her father's lifestyle choices. Ultimately, after learning from the young physician who has replaced the dead Doctor Moor that she will die from an unspecified terminal illness, Johanna burns her father's diary, locates a pistol in her home, and shoots herself in the heart.

Valhalla, structurally and substantively, draws on the multileveled meanings of "apocalyptic," including "revelation" and "cataclysmic end." As if to deliver an explicit reprimand, the narrator observes, "If John had any god it was his ax, / A very active god worshipped in use, / . . . Dividing the dead tree from the living forest. / It cut and split more wood than winter burned / And kept more pasture clear than there was need" (6.1871–1877). Too late, John (the Anti-Revelator, or "Reveal Later") admits to Johanna, "We see ourselves too clearly and the world / Not clearly enough" (6.1922–23). After reading John's diary, Johanna receives a vision: "She saw . . . his need for a world / Clean-cut, uncompromising, self-contained" (6.2152–53). Here, the narrator reveals to readers that the extinction of John's family results from human habits of wasteful excess. The moral: to force nature to become your heaven is to manufacture hell. Instead of descending suddenly in flood or famine, nature's inviolate judgment sweeps John's family from the earth with a silent decree of slow decades.[8] Iconic, repeated actions—such as when Johanna and young Judd visit the stone ruins of previous farms, and when Judd continues to pick, eat, and spit out sour crabapples—remind readers that nature's Eden has cast out, and will continue to cast out, human generations that partake of its fruit without replenishing the resources they have consumed (6.2050–55, 2165–71).

Structurally, Francis's six-part arrangement moves readers from what they expect—an anthropocentric division of four parts, à la James Thomson's *The Seasons*—and instead utilizes a messier, more environmentally accurate six-part structure that represents half-seasons, intermediate periods, and transitions that accompany traditional seasonal changes. The parts may also allude to the six seals referred to in the book of Revelation, the first six that precede the opening of the seventh and last apocalyptic seal.[9] It is also possible that Francis's structure is another attempt to displace the egocentric Euro-American ethos of his main characters and readers with a more ancient or Eastern paradigm, such as the Sanskrit *rtu-samhāra*, attributed to Kālidāsa, in which one calendar year is divided into six seasons.

Not without its flaws, where it lapses into anthropomorphic didacticism and simplistic fable, *Valhalla* still reads like a text ahead of its time—a gritty prophecy that adds its unbending realism and resounding clarity to a tribe of twentieth- and twenty-first-century works that form a choral warning. Its literary muscle resides in its complexity, the textual intricacies, and stylistic strata that match the endless and delicate interconnectivities of its subject. At times, the apocalyptic warning booms from the pages, although Francis's characters, and perhaps his readers, do not recognize or heed the earth's cataclysmic admonitions. When Leif and his sisters frolic naked in the thunderstorm, they do not recognize that the atmospheric doomsday symphony backing their revelry doubles as the final tumultuous notice that their generational blasphemy will soon be repaid: "Behind them when they turned / They saw the hill was blind with rain and lightning / . . . Thunder was mountains splitting and granite hales / Of mountains grinding together. They loved it loud" (3.966–70). When Johanna discovers John dead in the woods, the subterranean rumblings of the earth's tectonic tremors emerge as both speaking heart attack and earthquake: "Before she could run to him or move at all / She heard it coming up through the solid earth, / Pounding, pounding louder and louder the doom / Until Valhalla and all the mountains round it / Thundered. The end. The end" (6.1997–2002). Mystically, this apocalyptic moment occurs where the text's line numbering signifies the transition from one millennium to the next.

As if to level the playing field, in many places Francis entrusts the most dramatic scenes and lines to the non-human actors, which serves strategically to weave comparative foils through the zeniths of human drama.[10] When Leif lies unconscious, perhaps near death, on the forest floor after Judd's bare-fisted drubbing, Francis's description depicts the supreme indifference of the wilderness, his point being that human suffering is no greater than the suffering of the earth: "The trees stood round. / They murmured, but their murmuring said nothing. / They trembled, but their trembling was the air / In motion. . . . / They didn't care. The birds cared even less. / They didn't stop their feeding for a moment. / They flitted from bough to bough or flew away" (4.1263–70). Likewise, when the narrative lowers its final curtain of sky, it is not a human character who delivers the final soliloquy:

> She pressed the thing against her heart and fired.

> Angry, a crow dislodged itself from a tree
> And flew away. A thin blue dragonfly
> On a stem of grass had darted, circled, returned.
> Over the ground the ants kept on with their business
> Undisturbed. The air moved just enough
> To take another leaf, and then was still.
>
> $$(6.2335-41)$$

In a single stroke life changes completely, and life doesn't change at all. Everything has ended, and everything continues to begin.[11]

In *The Trouble with Francis*, Francis includes a rather tumultuous biographical vignette whose significance invites reconsideration in the company of a discussion about American apocalyptic literature. Two years before his move to Fort Juniper, Francis visited the MacDowell Colony for a third time, the occasion being the dedication of a plaque on a cottage occupied by Edwin Arlington Robinson. "Rain came that night and continued the next day as I drove home," Francis recollects. "It continued for five days and at the end was raining harder than at the beginning and with wind from the east" (*Trouble with Francis*, 20). The following scene conveys Francis's view from his "old house by the brook" in which he lived for three years without electricity or running water:

> Standing at a kitchen window and holding in place a piece of cardboard where one pane was missing, I began little by little to be aware that something unusual was going on. The wind picked up the rain and hurled it past in great sheets. But it was not until a stand of tall magnificent white pines nearby began to snap in two that I realized the force of the wind. I saw every one of those trees snap off, some nearer the ground, others higher up. I not only saw them go, I heard them; and later when I ventured outdoors I could smell them. The air smelled like a sawmill. It was September 21, 1938, the scheduled publication date of my book of poems, *Valhalla*. (*Trouble with Francis*, 21)

As fate would have it, one apocalypse invoked another. In this vivid self-portrait there is something enduring, a glimpse of Francis's essential character—a gesture both futile and mighty in the face of nature's awesome, destructive, and perhaps corrective wrath. In life, in poetry—what, after all, should the American ecopoet do but stand at the cracked window of heaven and hell and hold back a hurricane with his hand?

9
Conclusion

WHEN ONE STUDIES A WRITER'S LIFE, ONE IS called upon to examine one's own. In a recent issue of *Poetry*, John Barr notes, "As a Zen tea master, long before the ceremony of making tea, prepares the garden for his guests, sweeps the walk, cleans and composes the room, so poets should give their first attention to the lives they lead. . . . If they do not, how can poetry be a moral act? . . . Poets should live broadly, then write boldly" (Barr, 438). In Francis's case, perhaps "locally," "deliberately," and "intensely" could be added to Barr's observation, though whichever words we choose, there remains no doubt that, to Francis, lifestyle and poetic stylistics sprouted from the same taproot.

When Elinor Phillips-Cubbage interviewed Francis in 1975 at his Cowles Lane apartment in Amherst, prior to his final return to Fort Juniper, she noticed first the way the aging poet had arranged his surroundings. With obvious admiration and wonder, she noted that Francis, at the age of seventy-four, was still committed to outdoor walking "in all directions" along what he referred to as "fascinating routes" (Phillips-Cubbage, 153). Perceptively, Phillips-Cubbage identified the way Francis's temporary quarters fostered a life-giving mixture of inner and outdoor environments: a "Sunset Room" and six large windows like "living pictures set against plain white walls" that allowed him to see everything "from mountain ranges to Emily Dickinson's grave site" (Phillips-Cubbage, 154). When she asked Francis if the poet's life and work are "inseparable," he replied, "Although a poet or prose writer might try to escape from his life, he might not entirely escape. . . . I want to face myself as I want to face the outer world" (Phillips-Cubbage, 182). At one point, with sedate satisfaction, Francis told Phillips-Cubbage, "I have gone and lived other places. I enjoy variety. I'm not the sort of person who demands and requires a certain landscape" (156). A *certain* landscape, no. But *landscapes*, for certain. To live so that the landscapes of one's writing are always in view of the earth's landscapes, to sleep and wake in the presence of the rising and setting sun, even as one's writing swells and diminishes and one's aging body retrogresses toward the earth's elements—such artistic and natural alignments ordered Francis's daily movements and the movements of his daily writing.

But his journey was not always pleasant. That his marriage between living and writing became at times tenuous, almost threatened, can be divined from a terse journal entry dated June 4, 1945: "So far as I know, I am free to go on living here at Fort Juniper as long as I go on living" (*Travelling in Amherst,* 61). Five years later, during a sixteen-year publishing dearth that precipitated feelings of extremism and depression, Francis faced a personal crossroads at which his life and life's calling had begun to diverge. This "low point," as he called it, between 1951 and 1952—the period during which he wrote *Traveling in Concord*—pushed him to new levels of endurance. Wesley McNair writes that at this time Francis felt "traumatized by the literary world's disregard," to the extent that Francis subjected himself to voluntary incarceration in his bedroom with a drape covering its one window and a rug blocking the crack below the door, as if he were ashamed to come in contact with the light of day (McNair, 122). Francis remembers that he saw himself at this time contained within "three concentric strategic lines," the outermost representing his publication efforts, and the second line representing his writing. The third line, the one closest to him, had nothing to do with publication but was all about "living." "I might fail to publish poetry," he recalls saying to himself, "I might even fail in writing it, but nothing—if I passionately desired it—could make me fail in living it" (*Trouble with Francis,* 84).

Many of his poems center on this survivalist capacity to endure. His free-verse lyric "Monadnock" stands as a biographical allegory for its author's triumphant resistance, a characteristic he borrowed from his enduring mentor, the earth itself:

> If to the taunting peneplain the peak
> Is standpat, relic, anachronism,
> Fossil, the peak can stand the taunt.
>
> There was a time the peak was not a peak
> But granite and resistant core,
> Something that refused to wear
>
> Away when time and wind and rivers wore
> The rest away. Here is the thing
> The nervous rivers left behind.
>
> Endurance is the word, not exaltation.
> Two words: endurance, exaltation.
> Out of endurance, exaltation.

Francis's inspiration for this poem derived partly from lectures he attended at Massachusetts State College when he returned to school, in his forties, to study geology and botany. His lecture notes, complete with sketches, still survive in his archives. That he continued to write about the earth's eroded but lofty summits from his pauper's porch in life's valley reveals that the lessons he learned from the land provided life for his poetry. What Francis discovered then—and what he offers us now—is the map for a slow pilgrimage toward internal and external emancipation. His one-man quest for freedom from cultural constraints awarded him a personal solidarity that few people obtain. In rejecting materialism, he inherited an endless wealth of natural resources. By rarely stirring from his home in the woods, he traveled the universe.

As I continued to broaden my research and reread Francis's works, I became aware, gradually, of the spiritual poverty of my own suburban comfort in the twenty-first century. The more I wrote about Francis, the more I became dissatisfied with the wasteful, consumerist aspects of my life, the habits of excess and insensitivity toward the natural world that I had accepted as normal. Along the way, I was delighted to discover that individuals far more influential than myself, mostly among Francis's friends and fellow writers, had also found themselves nudged out of their lazy orbits in similar ways. Jean Loiseau, whose 1962 thesis was written to fulfill the requirements of the diplôma d' etudes supérieures at Bordeaux, remarked on the way Francis's writing attacked "a nation of household comfort, air conditioned, electrically heated . . . with a longing for wild open air life" (Loiseau, 85). Charles Sides, who interviewed Francis in November 1975, came away blissfully enlightened. "After our visit," Sides writes, "I felt that I had met a man capable of stilling the environment. Even today the sense of complete peace is my first impression concerning that visit" (Sides, "Freedom," 3). In addition, Sides describes his feelings of being "impressed by a unity" during his stay, a "feeling of quiet self containment that emanat[ed] from Francis and flow[ed] into the surroundings" (Sides, "Freedom," i). Likewise, Robert Bly found himself in awe of Francis's power to be "tuned in to the one inside" (Bly, 102). There is even some indication that future laureate Donald Hall, having admired Francis "for many years" and having read Francis's autobiography and *Collected Poems* with "greatest delight," learned of Francis's exemplary hermitage on the outskirts of Amherst and, following a time when he and Francis "breakfasted" together in Ann Arbor, as a result made his decision to forsake teaching and embark on his well-publicized country habitation at his ancestral Eagle Pond Farm in Danbury, New Hampshire (Hall, letter to Francis).

Without apology, Hall held up the senior Francis as a steady light to a generation of jaded poets and poetry readers who had lost their way. Over fifty years prior to his appointment as U.S. Poet Laureate, Hall befriended Francis, substituted for him at the Chautauqua Writers Workshop, solicited his poetry for anthologies and journals, and wrote no-nonsense appraisals of Francis's work in *Yankee* and *Ohio Review*. Hall's *Ohio Review* article—"Two Poets Named Robert"—praises Francis's refusal to ape confessional-type literary trends, his brevity, and his "close observation of things of the world, and the willingness to take pleasure in them" (Hall, "Two Poets," 116). Hall's candid assessment of the "Industry's neglect" of Francis deserves consideration again in the twenty-first century, as the industrial world and the poetry industry square off in an age of environmental anxiety and apathy. In short, Hall argues it was "just as well" that Francis remained largely unread, unpublished, and unrewarded during his life. He extols Francis's pacifism and vegetarianism and credits Francis's solitude with preventing him from "setting up a machinegun nest at Fort Juniper" and stopping him from defending his "privacy with a gun, as Gary Snyder . . . learned to do." Overall, Hall claims that Francis "survived intact, unflattered by fools" (Hall, "Two Poets," 125). Revisited today, Hall's incisive retrospective sums up the situation with clarity and accuracy: Francis hunkered down among those who walked tall through a corrupt and self-absorbed literary age, and he did so largely by substituting organic nature for human flattery.

Neither did Francis estivate nor hibernate too far afield to miss crossing paths with Marianne Moore. A quiet moment of correspondence of interest to modernists and ecocritics alike occurred between them on May 11, 1956. Moore wrote this letter to Francis from her Cumberland Street residence in Brooklyn on stationery from Amherst's Lord Jeffrey Inn. She sent the letter "in care of Mr. James Merrill" of the Department of English at Amherst College. Moore's letter arrived in response to the freshly cut woodland sprig Francis gave her as a going-away present in a custom-made "weightless box" of paper wrapped with "absolute exactness." "You have brought me a poem," Moore writes, "this flawless arbutus with its strange fragrance of the woods, and flawless leaves (how rare)." She narrates the remainder of her return to Brooklyn with dramatic delicacy, including her efforts to keep the wild cutting alive: "I thought I must not disturb it. . . . Then I thought about moisture, so opened the box. The moss was still wet, but I added water. . . . I know I shall reach Brooklyn with my sylvan treasure alive and beautiful. The varied tones of pink are a study, are they not?" (Moore, letter to Francis).

In a letter written that same day to Hildegarde Watson, Moore describes to Watson how Francis's gesture moved her: "Robert Francis,—a nature poet Macmillan has published several times—overwhelmed me by having gone into the woods in the rain and managed to gather enough arbutus to make a small bouquet, had wound wet moss round the stems, & tinfoil and put the treasure in a cracker box encased with great exactness in tissue paper,—all so painstaking and accurate!" (Moore, *Selected Letters*, 527). A decade later, in correspondence with the University of

Massachusetts Press's editorial board, Moore recalled this seemingly insignificant event in detail that suggests Francis's fusion of poetry and New England wilderness had a lasting impact on her (Moore, letter to Barron). All in all, Moore's interest in Francis went beyond a single thank-you note and confirms one major modernist poet's awareness of this obscure writer's importance to the world of poetry and the poetry of the world.

In fact, history shows that Moore acted as one of Francis's staunchest advocates. She urged Macmillan to publish *Selected Poems of Robert Francis: 1936–1950*. Editor Emile Capouya agreed. But the collection never appeared because, midway, Herbert Weinstock assumed Capouya's duties and nixed the project. Throughout the 1960s, Moore sent encouragement to editors who considered Francis's work. To the editorial board at Wesleyan University Press, who published *The Orb Weaver*, she wrote of Francis's "rare writing," described his poems as "souvenirs [that] have been lived," and counseled all publishers that "Robert Francis is (and would be) an ornament to any press" (Moore, letter to Wesleyan University Press). These and other remaining scraps of correspondence attribute Francis's modest resurgence into the publishing world largely to Moore's assistance.

In time, Moore's persistent support answered Macmillan's cold caprice. Fifteen years after Macmillan rejected the proposed *Selected Poems* title for Francis, the University of Massachusetts Press published Francis's *Come out into the Sun: Poems New and Selected* in 1965. Moore blurbed this book as "a beautiful object." "He is so penetrating, delicate and wise," she observed, "one goes away without having said a word, merely grateful to have received so much" (Moore, letter to University of Massachusetts Press). Later, Moore served as an advisory editor for the publication of Francis's prose collection, *The Satirical Rogue on Poetry*, though she disliked the title. When asked to help, Moore responded, "I am sure that anything that he wrote would have interest for me" and added that his book constituted a "superlative American product" (Moore, letter to Barron). During that same time, when interviewed by Muriel Rukeyser, Moore classed Francis as a writer "for whom she has affection" and "for whom she cares" (Rukeyser, 25). Though Moore's endorsements outweigh those of other supporters, she is not the only one who gravitated to the pulse of Francis's poetic fire.

As fate would dictate, Francis attracted the attention and company of many twentieth-century literary lions. Despite these brushes with greatness, however, he chose to remain obscure to protect his cherished solitude and independence. A small cardstock seating chart that survives in the Syracuse archives reveals that, in 1957, previous to his departure for Rome, the American Academy of Arts and Letters awards ceremony included Francis in the company of places reserved for Faulkner, Moore, Sandburg, O'Hara, Bishop, McCullers, Cheever, Porter, and Robert Lowell among others. Records indicate that, previously, Francis had judged Lowell (then at Kenyon College) above Louise Bogan and John Holmes in the Glascock Poetry Contest. At various stages in his life, Francis exchanged correspondence with May Sarton, James Dickey, James Merrill, Carolyn Kizer (who requested poetry from him while editing *Poetry Northwest*), and Richard Wilbur. Dickey called *The Orb Weaver* "damned fine," said the poems made him "live," and requested from Francis a "good long talk . . . as long as the trees themselves" (Dickey, letter to Francis). According to Fran Quinn, at the memorial service held for Francis at Amherst's Jones Library, Richard Wilbur remarked that *The Orb Weaver* was the finest book that Wesleyan University Press had ever published (Quinn, telephone interview with author, December 2008). James Merrill dubbed Francis's poems "all irresistible" and admired most the "truth" of their "vision." In one letter, Merrill sent Francis a poem in draft—"The Sneer," presumably for feedback—and mentioned that he kept Francis's poem "The Brass Candlestick" on his mantelpiece near a candlestick (Merrill, letter to Francis). Francis's pristine archival material delivers a fragmentary narration of these and other connections with literary giants of his age. Because he cherished his seclusion, though, he rhapsodized on coming out into the sun while remaining out of the limelight.

Though it is difficult to put into words, I can only say that the changes that Francis's literary legacy have wrought in my life—as a teacher, reader, writer, husband, father, and citizen of the world—have been pervasive. And they are still ongoing. At the outset of this project, I wondered what someone from the semi-arid plateaus of eastern Idaho's Snake River region could have to say about a poet from New England. My answer to this question is that Francis's poetry achieved a certain wide range of appeal. The meditative qualities he found in Amherst's twentieth-century fields, forests, and streams can be transferred and understood vicariously by others who read his work but who may inhabit extremely different twenty-first-century bioregions—alpine, coastal, desert, urban, rural. Francis's poetry makes possible a feeling of inner transcendence, personal and environmental awareness, and peaceful reflection that passes for travel around the globe. His writing acts as a portal for what is geologically and geographically analogous, or "geologous," between cultures, species, populations, and landforms across time and space. His poetry reveals how divergent landscapes and bioregions, when the poems written about them produce similar reactions in readers, act as "geologues" for each other. Why else could my reaction to his art have been so arresting? How else could I have read "New England Mind"—while looking westward out my kitchen window at Menan Butte rising from the sagebrush-dusted expanse of lichen-scorched lava rock—and understood immediately what Francis meant?

When I learned that no one had yet published a full critical study of Francis's work, my task turned quickly from academic drudgery to harried quest. And I have to admit—though it might sound hokey—that there were times when I felt as if Francis were speaking to me from the archives and databases, urging me to get the job done. While reading his 1978 *New England Review* article "Frost as Apple Peeler," online, for example, I came across a line where Francis appears to forecast the possibility of future scholarly research on his life. "It has been read by I don't know how many people," he writes—referring to Frost's cruelly homophobic poem, "On the Question of an Old Man's Feeling"—"and may be for all I know accessible to scholars in the future" ("Frost as Apple Peeler," 38). At the time I read this, I had quite literally requested a copy of Frost's anti-gay diatribe from the Dartmouth archives and had just received it in my office mailbox. Another moment: On May 10, 2007, I sat on the upper floor of the Bird Library at Syracuse University, puzzling over the source of a heady aroma that had saturated the rainy air on campus for days. And the first piece I pulled from the archives was "Thoreau's Search," a *Christian Science Monitor* essay that Francis wrote about Thoreau's seven-year search (also in May) for the source of a wilderness fragrance he couldn't identify. The date of the essay was May 10, 1939. The serendipitous coincidences continued when, after reading Francis's short essay "A Golden Simplicity?" I got the eerie feeling that Francis was somehow describing the way I had discovered him, narrating, as it were, the story of our union before it happened. "A Golden Simplicity?" describes the recreational and desultory habit of reading through anthologies—"just to see what is there"—as I had done when I found Francis. "Your eye is caught by something in the poem before you, some slight movement or the hint of it," the speaker explains. "And before you know it the poem of its own accord rises from the page and comes toward you!" (*Pot Shots* 77). This is how I found Francis in the Longman anthology at the Idaho Youth Ranch. In the same way that I have found my favorite landscapes, I found the poet—precisely when I wasn't looking for him.

While completing my coursework in the doctoral program at Indiana University of Pennsylvania, I began to set apart a few hours on weekends for hikes at Yellow Creek State Park, my sole purpose being to increase my knowledge of wild bird, animal, insect, and plant species, as well as to soak in more outdoor air and sunlight. My most rewarding mini-retreats started from an access road, through some Amish farms, on the southern side of the reservoir. On my hikes to the dam site, I saw kingfishers that bobbed on tree limbs, then zipped toward the flashing water in streaks of emerald lightning. I was chastised by squirrels and blue jays and bombarded by kamikaze needle-nosed insects of all kinds. For days, before I discovered the source, I was baffled by the baritone recitatives of herons, and once, while sitting silently near a stream, I ducked when from above a flapping shadow swooped down as if to sink its mythical roc's talons through my clavicle and carry me aloft. But when I glanced at the limb overhead, I saw a pileated woodpecker as big as a rooster. Oblivious to my presence, he cocked his red beret back on his head, shook the dust from his salt-and-pepper mantle, and set about scaling the rugged gray bark of an ancient maple. As he ascended, I listened to the skittering clatter and snatch of his feet against the bark. He worked his hooks against the trunk with the dexterity of a rock climber. His pickaxe beak struck gold in every crevasse and groove.

Below the dam I found a deep eddy and a natural parapet of gray granite. At this gentle bend in the river, the opaque water slowed and swam with turgid, sulfurous runoff and copper sun-glare. As often as I could (I have forgotten how many times I went), I would swim and sunbathe there in an effort to clean from my skin the sticky summer Pennsylvania heat and humidity. Though secluded, I could tell that others had found my Eden first. Shimmering tangles of rusted fishhooks, fishing line, and red-and-white bobbers trailed their redundant rhetoric through the smooth multicolored river rocks. Charred heaps of ashen tree limbs and beer cans peppered the riverbank like landmines.

Still, amid the evidence of human travesty, I saw unabated growth and renewal. Countless canoe birch stubs, their tops having been gnawed to blunt points by beaver, rose from the black humus like pickets and more than once tripped me up as I walked along the bank, eyes skyward. On the forest floor, I saw something I had never seen before. Littered everywhere, among the roots and weeds and flowers, appeared to be hundreds of green lace doilies—the "understory," or next generation of coniferous evergreens, driving toward the sun on a seventy-year slow-motion trajectory.

On each trek out and back, I passed a tree that had grown rather impertinently in the middle of the path. In August, on my way back to my car, I halted at the familiar tree and stared, dumbfounded, because somehow it had grown apples. While stretched on the rocks to dry off (being careful to dodge the spiders!), I watched a muskrat trundle its cargo across a natural bridge of toppled birches, and I held my breath as a doe as red as cinnamon on the opposite bank nosed her way through the decaying leaves and disappeared into the swaying cathedral of trees. Eventually I was able to identify the beebalm's hot-pink sunburst, the eastern black swallowtail butterfly, the lake perch, and the whirligig water beetle. With the aid of a library field guide, I spotted the polydore cinnabar growing like a brilliant orange saucer from the side of a massive oak, a rusty discus slammed into a pillar of ancient wood.

Without a doubt, though, my most memorable catch occurred when, in the middle of dozing and sunbathing, I heard a scream, opened my eyes, and scanned the sky. Swirling high above the trees, like angry acrobats, two red-tailed hawks slashed and grappled at each other in a dogfight over a fish. At the exact moment I caught sight of them, one hawk dropped the fish, wheeled away, and the other swooped in and snagged it in mid-air. It was a sight I had never witnessed before and have not witnessed since. At the time, the sound that jolted my heart was the scream. Pure rage, a searing shriek, a sound both pure and painful—the scream of the raptor defies description.

But that day I was in the unique position of having read Francis's poetic descriptions of hawks before seeing these two in the wild, and the occasion provided a memorable moment in which art fused with life. "Who is the hawk whose squeal / Is like the shivering sound / Of a too tightly wound / Child's toy that slips a reel?" he asks in "The Hawk" (1–4). And in "Demonstration," Francis describes the visual and aural artistry of the hawk: "With what economy, what indolent control / The hawk lies on the delicate air, looking below / He does not climb—watch him—he does not need to climb. / The same invisible shaft that lifts the cumulus / Lifts him, Lifts him to any altitude he wills. / Never his wings, only his scream, disturbs the noon stillness" (1–6). It has been this poetic quality that has changed me the most—Francis's fusion of literature and the earth's phenomena, accompanied by an opening of the eyes and a desire to observe rather than to hurry by in my pursuit of understanding and meaning.

There were times, too—such as the present moment—when I wondered if I was doing Francis more harm than good by making him the subject of intellectual and academic inquiry, since for nearly fifty years he had reigned at the anti-university of Fort Juniper as one of the world's most notorious but obscure iconoclasts of overweening intellectualism. In many ways, his retreat to the woods outside Amherst was an escape from the malaise of the rationale and the mechanized depravity of industrialized suburbia. The "private fire" he wrote of in "Three Darks Come Down Together" may have been the one he kept burning himself, inside his poetry, a torch to ward off the ever-nearing approach of evil, the ravages of a domineering intellect, and the carnassial fang of the human hunger to develop, subdue, and tame. Still, over twenty years after his death, in the relative absence of any serious more global resurgence of interest in his work, I am confident that this re-introduction is better than none, particularly at a time in the history of our planet when his message is so sorely needed. The private fire that Francis kindled will soon, I hope, catch the wind and sweep with purifying swiftness elsewhere.

If this study marks the start of a glimpse into the future of Francis's legacy, we should pause to consider that his world was the past world. The anachronism he mentions in "Monadnock" was in many ways himself—the immovable peak, the poet that time couldn't wear away. If Aldo Leopold declared the virtues of "thinking like a mountain" (Leopold, 132), Francis took one giant metaphysical step beyond that to say we should come as close as possible to *being the mountain*. Everywhere in his writing, we find evidence that he felt out of sorts with the time in which he'd been born. He found himself less and less drawn toward his contemporaries, and more at home in the company of nineteenth-century authors. "Possibly the only shocking confession I have to make in this book is that I do not like poetry," he unapologetically ends his autobiography. Contemporary poetry at its best bores me, and it is not often at its best. . . . To visit a poet is to put oneself at the mercy both of his poetry and of himself. While he would be having a hell of a good time drinking his beer, smoking his pipe, and spouting his poems, I would be without the beer, without the pipe, and praying to God to be without the poems" (*Trouble with Francis,* 131).

Early on, and all through his life, he grappled with the paradox of poetry: the art that he loved seemed to drive him from the life he loved. So he wrote and remained alone.

He was so unlike poets of his time, so similar to poets of the past. Like Dickinson, he stayed home—and visited the universe. Like Thoreau, he built his home in the woods. But unlike his famous predecessor, Francis stayed. Like Emerson—whom he addressed as his "master" in a 1937 journal entry—he lived nearly from the beginning to the end of his century, fell out with God, and found God again in himself and his surroundings. "I love this world of sky and ground and hill and growing things so dearly that I am a little trouble by my externality," he wrote in his journal. "I feel an urge to pierce through the familiar aspect, the habitual appearance, and to seize essences" (*Travelling in Amherst,* 3). Part of this piercing through the veneer of humanity required him to cast his sight backward, to emulate nature first-hand, and to slough off the straitjacket carapace of civilization. In his poem "Emergence" and in the preface to his *Collected Poems,* he describes this process by putting himself into the position of one of his favorite natural entities: the praying mantis. The metaphor of the molting mantis provided Francis with an ideal icon of ecosystemic identification, one that captured his life cycle completely. "It may amuse you," he writes, speaking of himself as the author of his collected poems, "to follow this constant coming out of himself, like a mantis in its successive moltings. Only in its final molt does the mantis achieve its wings, and here our analogy breaks down, for this poet, like all poets, does a little flying all along the way."

The last poem in Francis's *Collected Poems*, "Fire Chaconne," ignites an elemental echo to the private fires he lit in earlier poems such as "Three Darks Come Down Together." This sequence of haiku-like clusters, unlike anything Francis had written previously, is evidence that he continued to molt and stretch his wings to the end of his life. "Fire Chaconne," as Robert Shaw observes, consists of "twenty tiny poems all centered on the image of fire," its structure perhaps influenced by Wallace Stevens's "Thirteen Ways of Looking at a Blackbird" (Shaw, 87). Each ideographic couplet crackles on the coals of its own making. Number two: "Firefly, your green / Spark on the green darkness" (2–4). The sixth and seventh: "The cold stars / Their indelible heat. / Blue juniper / Berry: two summers' sun" (11–12). Nine: "Flaming in fall / Gorgeous the poison ivy" (17–18). And eighteen and nineteen: "Bonfire on snow / Twice beautiful: symbol and fact. / Almighty sun / Father of all our fires" (35–38). In the seventeenth couplet, he preserves his poetry's personal statement: "True poem / Burns on undiminishing fuel" (33–34). In the third section, his spin on the Cartesian dictum secures his unique place in American ecopoetics: "I burn / Therefore I am" (5–6).

Of all that has been said of Francis, and all that may yet be said, I am persuaded to agree with Donald Hall, who called him "a modern American classic" (Hall, "Two Poets" 123). I also agree with Richard Gillman, who called him "a student of nature and human beings, a philosopher and a man of independence" (Gillman, Intro., *Travelling in Amherst*, vii). But to me, Francis will always be the keeper of an enduring flame. In a world continually devastated by destruction and wanton disregard for nature, Francis's poetic fire burns nothing and nowhere but in the cavernous souls of those who wish to see the earth preserved. Though he didn't live to see the full scope of his literary legacy, the world he loved may still endure to witness his private fire circle the globe.

Notes

CHAPTER 1. INTRODUCTION

1. In *Painted Bride Quarterly*'s special Francis issue (1988), Fran Quinn indicates that Francis's "lack of recognition should be rectified" (Quinn, 5). Robert Shaw writes that Francis's "work has yet to receive much close critical attention" and correctly observes how most Francis criticism has "offered more appreciation than interpretation" (Shaw, 90). In *Encyclopedia of American Poetry*, Karen Stein claims, "It is time for a reappraisal and revaluation of Robert Francis" (Stein, 222). In her master's thesis, "Robert Francis: Best Neglected Poet," Dolores Whitney concludes, "It is my hope that lovers of modern poetry will delve further into the works of Robert Francis . . . A further study of [his] nature poetry is called for" (Whitney, 81–82). In 1981, David Young noted the "troublesome fact" that Francis, at the age of 80, was "so little known" (62).

2. See Stephen Dunning's article (co-authored by Francis), "Poetry as (Disciplined) Play" in the *English Journal* (1963). Dunning writes, "How fine a day that would be if everybody in your class knew as much about poetry as he does, say, about baseball" (Dunning, 601). As recently as 2006, a student at Midlands Technical College in South Carolina indicated in an email that his instructor used "Catch" as the inaugural poem for his entire course in poetry (email to author, April 14, 2006).

3. In his interview with Elinor Phillips-Cubbage, Francis reveals that "The Pitcher" functions as a metaphor for the relationship between "an author and his reader" (Phillips-Cubbage, 160). Also, Wesley McNair, in *Mapping the Heart*, reads "The Pitcher" as a play-by-play commentary on "art": "We might not see at first (temporarily blinded, like the batter) that the pitcher is also a poet, and that what has gotten by us is a description of how the poet works. Look again at the pitcher's windup, . . . and we see the pitcher is very much like Francis himself" (McNair, 118–19).

4. David Kelly's entry on Francis in the Gale Group's *Poetry for Students* volume uses a single poem, "The Base Stealer," to classify Francis as "one of the best baseball poets ever" (Kelly, 35). Steven Rizzo's pitcherly pun in his 2007 *Explicator* article places Francis "well beyond the minor leagues of poetry" (Rizzo, 114). Author Michael Ceraolo's poem, "The Base-Stealer Reprised," recently appeared in *Slow Trains* and alludes directly to Francis's poem. In an email, Ceraolo reported that he discovered Francis's work in "an anthology of baseball poetry" (August 22, 2008).

5. In 1976, Rudy Kikel reviewed Francis's *A Certain Distance*, whose "Apollonian" and "*passive* Dionysian" extremes of "Platonic abstraction" and "erotic inundation" Kikel contrasted to the "formless" poems that "flood gay quarterlies."

6. Elinor Phillips-Cubbage applies the moniker "minority man" to Francis (Phillips-Cubbage, 149), and Charles Sides notes that Francis often addressed "himself as a poet, minor" (Sides, "Freedom," 60). Robert Wallace observes in Francis's poetry a "minor, hidden exactness" (Wallace, 16), and David Graham writes, "[C]onsider how little he fits our common image of a 'major' poet. Despite the admitted absurdity of the terms 'major' and 'minor' . . . the critical consensus on Francis has often seemed to be that he is a fine but minor lyricist" (Graham, 81, 83).

7. In a letter to Louis Untermeyer, Frost attacks "the majority and minority opinions" on Francis's work that Untermeyer had "handed down" (Phillips-Cubbage, 167). Mary Fell asserts, "Robert lived a marginal existence both financially and socially. Had he not achieved self-sufficiency as perfectly as he did with the Fort, he might have remained a marginal man, a local eccentric tolerated on the fringes of Amherst life but never able to enter its center. He would have had to live on others' terms as best he could, instead of creating his own" (71). Robert Shaw offers, "The consensus, unlikely to be changed, is that Frost is a major poet, Francis a minor one. . . . Without disputing the label, one might note that Francis, unlike many minor poets, was not only keenly aware of his limitations but was eventually able . . . to turn them into strengths. The individualism of his outlook, the clarity of his perceptions, and the consistently high quality of his unshowy craftsmanship have few parallels in modern American poetry. His poems continue to remind his readers that the pleasures afforded by some minor poetry are anything but minor" (91).

8. Andrew Stambuk's "Learning to Hover: Robert Frost, Robert Francis, and the Poetry of Detached Engagement" includes Louis Untermeyer's scathing "animadver[sion]" in which Untermeyer claims that Francis's "The Wood Peewee," "The Wasp," and "Statement" read "like poems Frost had been writing and not yet decided to print," adding, "They are admirably neat, they are playfully philosophical, they blend observation with imagination. But we know who wrote them first" (Stambuk, 535).

9. Stambuk assembles a cogent cross-textual examination of titles and topics to which we could add others. For example, Frost: "Mountain Blueberries," "The Oven Bird," "Design" (spider sonnet), "Come In," "To Earthward," "The Figure in the Doorway," "Mending Wall," "Two Tramps in Mud Time," "Nothing Gold Can Stay," "An Encounter," and "Fragmentary Blue"; Francis: "Blueberries," "The Wood Peewee," "The Orb Weaver" (sonnet-like spider poem), "Come Out into the Sun," "Homeward," "The Face Against the Glass," "The Wall," "Two Bums Walk out of Eden," "The Name of Gold," "Encounter," and "Exclusive Blue." Robert Shaw's *American Writers* entry on Francis labels this haunting type of influence "almost one of ventriloquism" (Shawn, 76).

10. Wesley McNair classes Francis's "revolt against Frost" as "one of the most interesting aspects" of Francis's career. McNair argues that the "story of how Francis freed himself, . . . overthrowing his mentor Robert Frost as he did so, is enough to encourage any aspiring poet" (McNair, 119, 129). Andrew Stambuk labels Francis's poem "For the Ghost of Robert Frost" a kind of exorcism that indicates "Francis took Frost's advice more to heart than Frost ever did" (Stambuk, 551). David Graham argues that after *Valhalla and Other Poems* in 1938 Francis "mostly succeeds in shaking off the oppressive echoes of Frost" (Graham, 84).

11. In *Ecopoetry: A Critical Introduction* and *The West Side of Any Mountain: Place, Space, and Ecopoetry*, J. Scott Bryson defines ecopoetry's three characteristics: "an ecological and biocentric perspective recognizing the interdependent nature of the world; a deep humility with regard to our relationships with human and nonhuman nature; and an intense skepticism toward hyperrationality, a skepticism that usually leads to condemnation of an overtechnologized modern world and a warning concerning the very real potential for ecological catastrophe" (Bryson, *The West Side of Any Mountain*, 2).

CHAPTER 2. DICKINSON AND FROST

1. During the year that Francis taught at Mount Holyoke, Ulrich Troubetzkoy noted in Francis's poems certain "capsules of philosophy that remind one of Emily

Dickinson" (Troubetzkoy, 26). Howard Nelson contrasts the way Francis "rarely stops the reader in his tracks" to the way Dickinson "often stops her reader dead every line or so" (Nelson, 4). David Graham argues that the "shade of Emily Dickinson" prevents future readers from calling Francis even "the second-best poet to write in Amherst, Massachusetts" (Graham, 86). After hearing Francis read, David Walker remembers how the experience, "in Dickinson's phrase, took the top of [his] head off" (Walker, "Francis Reading," 29). Fran Quinn attributes Francis's settlement in Amherst to Francis's desire to be closer to one of "two historic loves," the first being Dickinson (Quinn, 7). And Alan Sullivan compares phallic imagery in Francis's long poem *Valhalla* to "alleged clitoral references" in Dickinson's poetry, arguing that at least thirty of Francis's poems "compare well with anything" written by "Miss D." and that in specific Francis lyrics "the ghost of Dickinson seems to peer over his shoulder" (Sullivan, pars. 16, 18, 26).

2. Bloom's notion of influence finds the "later poet" writing in a way that represents "movement towards discontinuity with the precursor," toward a "range of being just beyond the precursor," and toward a state of being "so as to separate himself from . . . the precursor" (Bloom, 1803). While Bloom's model certainly applies to Francis's relationship with Frost, Francis appears to have welcomed—if not sought out—associations between himself and Dickinson. To Gilbert and Gubar, the female poet, fearful that the "act of writing will isolate or destroy her," is assaulted by a "radical fear" that prevents her from creating because "she can never become a 'precursor'" (Gilbert and Gubar, 2026). Francis's emergence as a poet appears to re-arrange the gender polarities in this instance, in that as a male poet he found even greater freedom to write in the shadow of Dickinson, his female predecessor and fellow resident of Amherst.

3. A must-read for Frost scholars, *Frost: A Time to Talk* is, as Robert Shaw perceptively labels it, "brief but absorbing," "respectful but not uncritical," and "balanced and without apparent prejudice" (Shaw, 85–86).

4. Following a tête-à-tête with Frost in 1933, Francis relates how he and Frost discussed the sway that Dickinson maintained over them, in some ways deflecting her power with criticism, in others acknowledging her as mistress and master: "Emily Dickinson is a poet whose 'state' never gets sidetracked. Since she wrote without thought of publication and was not under the necessity of revamping and polishing, it was easy for her to go right to the point and say precisely what she thought and felt. Her technical irregularities give her poems strength as if she were saying, 'Look out, Rhyme and Meter, here I come.' Frost likes this willfulness, this unmanageability of the thought by the form, but he thinks it was a little too easily arrived at by Emily Dickinson in whom it is sometimes indistinguishable from carelessness. In other words, she gave up the technical struggle too easily" (*Travelling in Amherst*, 31).

Chapter 3. Sex, Gender, and The Rural Erotic

1. In *De-centering Sexualities*, Richard Phillips and Diane Watt observe, "Much less has been said about other liminal or in-between spaces, including the small towns and rural parts of Europe, Australia, and North America. Yet these spaces may be of great significance, with respect to representation and politics of sexualities, for it is in such spaces that hegemonic sexualities may be least stable" (Phillips and Watt, 1).

2. In "Eroticizing the Rural," David Bell writes, "[T]he rural occupies a very particular, but very complex, location in the wider sociospatial economy of desire." This "archeology of sex in the 'middle of nowhere,'" he adds, "aims to think about some popular representations and performances of the erotic rural—of the links made in

the popular imagination between nature, rural life, and forms of sexual activity and identity" (Bell, 84).

3. Shuttleton argues that poems that employ a "queer pastoral imaginary" can serve both "elitist and emancipatory, aesthetic and polemical . . . structures of feeling." He adds, "Queer pastoral can be read as personally ennobling and culturally restorative on the one hand, whilst on the other repudiated as 'rural idiocy'" (Shuttleton, 129–30).

4. Jeff Morris labels *A Certain Distance* "an ill-conceived project" of "poem-pictures" characterized by "embarrassing sentimentality and simpering obliquity" (Morris, 59). Contrariwise, Alan Sullivan finds Francis's "statuesque . . . boy-poems" pleasantly "less guarded" because "the reader senses a terrible tug at a willed restraint" and because "Francis spoke as a gay man to a straight audience" and "loved beauty for its own sake, whether in boys or in verses" (Sullivan, par. 10). A *Small Moon* review classes the "aesthetically and philosophically pleasing young men" in *A Certain Distance* as "real, integrated people rather than merely objects," a collage of "sketches of the integrated soul moving through time . . . , of one man who is several men" (*Small Moon*). Rudy Kikel, however, in the *Advocate*, argues that the prose-poems provide a "taste" rather than a "full meal." Kikel suggests that the writing reveals Francis's desire for "a freedom from sexual complexity that the poet may have wished for himself." Ultimately, Kikel puts distance between himself and Francis's book. He asks, "Is Francis subtly homophobic?" (Kikel).

5. Graham elaborates: "For the swimmer, all contact is in a real sense only skin deep. . . . We may achieve a momentary illusion of such union, just as we may swim as if without effort, or float as though permanently at ease on the water's surface—but the water remains an alien element, and we can drown in it. . . . With this final image, Francis seems at first to describe an unambiguously sensual, affirmative vision of love-making, but the metaphor's implications emphasize solitude" (91–92).

6. In "His Running, My Running" David Graham sees a "profoundly, tenderly voy-euristic" and a "discreetly erotic yearning" that "recalls nothing so much as Section 11 ['Twenty-eight young men'] of Whitman's 'Song of Myself'" (89–90). More interested in facts than connotations and textual allusiveness, Alan Sullivan reads the biographical through the biological: ["His Running My Running"] tells the story of its author—a watcher of boys who has belatedly enjoyed his hour of sensual pleasure in the early autumn of life and now, as days grow short, sees that hour receding swiftly into the distance" (Sullivan, par. 44).

CHAPTER 4. FICTION AND NON-FICTION

1. Karla Armbruster assembles criteria for a text that qualifies as a bioregional narrative: (1) it presents "human activities and communities as compatible with the natural aspects of place"; (2) demonstrates "the ethics and practice of interacting sustainably with both human and natural communities"; (3) views nature as "one more player in the construction of community"; (4) emphasizes a "sense of reciprocity between human and nonhuman players"; and 5) projects an "ideal balance of nature and culture" (Armbruster, 9, 10, 13).

2. H. L. Varley savored its "Thoreau-like economy" (Varley, 2). An *Amherst Record* critic, "F. P. R.," correctly observed that "[n]ot much, fictionally speaking, happens" but that it departs refreshingly from the "usual fictional trumps—love, manslaughter, and mystery"—to explore the stultifying "pattern and paralysis" of provincial, middle-class New England life: "One enjoys Robert's becoming acquainted with himself and coming to terms with his environment" (F.P.R., 10). Richard Gillman blames the

"Atomic age or 'age of anxiety'" for driving readers toward "another Hemingway novel" instead of Francis's "calm and serene" narrative of the biocentric sublime (Gillman, "Amherst Poet").

3. Karla Armbruster, in reference to the bioregional narrative's "contingent nature," notes how such narratives "stress a more postmodern, decentered concept of self defined by relationships with both nonhuman and human others so often left out of traditional nature writing" (Armbruster, 10). Also, following its publication, Richard Gillman identified *We Fly Away*'s bold structural experimentation: "Mr. Francis does not believe in mechanical form in the respect that the author should pour his thoughts into a preconstructed mold. He firmly feels the composition should find its own form as naturally as a tree makes its own width, depth, and height" (Gillman, "Amherst Poet").

4. An anonymous reviewer—O. O.—described *We Fly Away* as "feather-delicate" and "ornithological-hearted." (. . .) Mitchell Thomashow lists several "perceptual guidelines" for ways to "link the bioregional and global, in order to construct a cosmopolitan bioregionalism," the first of which is "[s]*tudy the language of the birds*" (Thomashow, 130).

5. Bernard Quetchenbach cites 1948 as the year of "the birth of contemporary poetry," a time when "the relationship between the individual human life and the environment [went] from idiosyncratic hobby to a matter of clear societal importance" (Quetchenbach, 28).

6. Hoffman's first three categories for the American novel up to 1950: 1) "the novel of World War II" (Robert Lowry's *Casualty*, Mailer's *The Naked and the Dead*); (2) "the spectacle of the established novelist who continues to exploit those characteristics which originally helped to established his reputation" (Steinbeck's *Cannery Row*, *Tortilla Flat*, and *The Wayward Bus*, and Hemingway's *Across the River and Through the Trees*); and 3) "the persistence of naturalism" (Nelson Algren's *The Man with the Golden Arm* and Saul Bellow's *The Dangling Man* and *The Victim*) (Hoffman, 171).

7. Joseph Meeker observes, "The evolutionary process is one of adaptation and accommodation, with the various species exploring opportunistically their environments in search of a means to maintain their existence. . . . Successful participants in it are those who remain alive when circumstances change, not those who are best able to destroy competitors and enemies" (Meeker, 33, 35). James Perrin Warren's reading of Whitman uncovers a "theme of evolution" with which Whitman explores ways that "humankind will model itself on the landscape," a "combinatory vision" of writing and living that "crosses any scientific boundary between different species" (Warren, 168, 173).

CHAPTER 5. ECOSPIRITUALITY AND ECOPOLITICS

1. Aldo Leopold's *Sand County Almanac* introduces the concept of the "biotic community" in which "man" slips from "conqueror of the land-community to plain member of it." In contrast, exploitive ways of living in conjunction with the earth are labelled "Abrahamic" in reference to the Old Testament patriarch who viewed land as a means of dripping "milk and honey" into his mouth (Leopold, 411). Also, Lynn White Jr. famously argued: "The whole concept of the sacred grove is alien to Christianity and to the ethos of the West. . . . We shall continue to have a worsening ecologic crisis until we reject the Christian axiom that nature has no reason for existence save to serve man. . . . Both our present science and our present technology are so tinctured with orthodox Christian arrogance toward nature that no solution for our ecologic crisis can be expected from them alone" (White, 12, 14). Also, Daniel Spencer refers to work of Larry Rasmussen:

"What is most needed today is *conversion* to the earth—a turning to earth in both orientation and allegiance" (Spencer, 364).

2. Leonard Scigaj cites Matthew Fox to differentiate between pantheism and panentheism: "Fox prefers *panentheism* to *pantheism*, because pantheism, heretical to Christians, robs God of transcendence by equating his essence with his material creation—hence 'everything is God and God is everything.' Adding the green *en* results in the acceptable assertion that 'God is in everything and everything is in God'" (Scigaj, 120–121).

3. The King James Version of the Bible renders John 3:8 this way: "The wind bloweth where it listeth, and thou hearest the sound thereof, but canst not tell whence it cometh, and whither it goeth: so is every one that is born of the Spirit."

4. Andrew Stambuk writes that Francis's "The Orb Weaver," like Frost's "Design," is "suggestive of nature's malevolence." Like Frost, Stambuk adds, Francis "incorporates into his poetics a perception of nature that is quite different from the benign view he presents in many of his other poems" (Stambuk, 550).

5. Greg Garrard aligns "eco-politics" with the "poetics of responsibility" in order to claim that "we should not disguise political decisions about the kind of world we want in either the discredited objectivity of natural order nor the subjective mystification between nature and culture, its construction and reconstruction" (Garrard, 179). Ursula Heise writes of ecocriticism's focus on "political subjecthood" and its "triple allegiance" to the scientific study of nature, the scholarly analysis of cultural representations, and the political struggle for more sustainable ways of inhabiting the natural world" (Heise, 506). Daniel Spencer lists Douglas John Hall's four principles of environmental "stewardship," which include "*ecologization*," the moving of ethics "beyond the human community to include the stewardship of the earth and its creatures"; and "*politicization*," which "implies both moving stewardship out of the realm of sentimentality and private morality to address the difficult political realities of the worlds, as well as extricating stewardship from its associations with economic capitalism" (Spencer, 138).

CHAPTER 6. ECONOMY, PLACE, AND SPACE

1. Ursula Heise observes that ecocriticism aims its "critique of modernity at its presumption to know the natural world scientifically, to manipulate it technologically and exploit it economically" (Heise, 507). Lawrence Buell writes, "In western culture, the order of nature has been variously imagined . . . as an economy (from the Greek *oikos*, household). . . . The obsolete metaphor of nature as 'economy' has been revived in the twentieth century both by ecological science . . . and by environmentalists stressing the ancient understanding of economy as the manifestation of divine order in the form of local stewardship precisely in order to assail the industrial economy" (*Environmental Imagination*, 281, 283). In *The Future of Environmental Criticism*, Buell showcases "economy" as a keyword in his glossary and lists *oikos* as the "root of both 'ecology' and 'economy.'" This "provenance," Buell argues, "together with economy's originally cosmic implication, connoting the divinely appointed order of things, has prompted some modern environmental writers like Wendell Berry to call for a reorientation of the secular (or 'little') economy in accountability to the 'great' economy" (Future, 140).

2. J. Scott Bryson provides the following definitions of *place* and *space*: "Ecopoets offer a vision of the world that values the interaction between two interdependent and seemingly paradoxical desires, both of which are attempts to respond to the modern divorce between humanity and the rest of nature: (1) they *create place*, making a conscious and concerted effort to know the more-than-human world around us; and (2)

to *value space*, recognizing the extent to which that very world is ultimately unknow-able. . . . [T]he project undertaken by contemporary ecopoets falls somewhere within these two objectives, to know the world and to recognize its ultimate unknowability" (Bryson, 8).

3. In defining the relationship between place and space, Bryson cites Yi-Fu Tuan's *Space and Place: The Perspective of Experience*. Tuan writes that place and space "require each other for definition. From security to stability of place we are aware of the openness, freedom, and threat of space, and vice versa" (*West Side of Any Mountain*, 9). Further, Tuan situates all human lives in a "dialectical movement between shelter and venture, attachment and freedom," adding that in "open space one can become intensely aware of place; and in the solitude of a sheltered place the vastness of space beyond acquires a haunting presence." He adds, "A healthy being welcomes constraint and freedom, the boundedness of place and the exposure of space" (20).

4. In her exploration of Daphne Marlatt's poetry about the environment and the les-bian body, Beverly Curran notes Marlatt's similar strategic use of poetic point of view: "Marlatt's use of the first-person plural admits her own membership in the culture of consumerism and her complicity in the damaging of the environment, but it also seeks out her reader in an act of collaborative resistance" (Curran, 198).

Chapter 7. The Experimental Environmental

1. Laird Christensen writes, "Although postmodern theory and ecology may appear to be at odds with one another—one questioning the very notions of reality in which the other is grounded—it is useful to consider how these two ways of understanding the world are complementary" (Christensen, 135). Also, according to George Hart, "Postmodernist poetry does not find much acceptance or generate much interest among readers of nature poetry and ecologically oriented critics. . . . Nonetheless, some post-modernist poetry can make a valuable contribution to the green canon even as it turns to textuality, the materiality of the signifier, and experimental rhetoric and figurative language" (Hart, 315). Further, Gretchen Legler's exploration of the "postmodern pas-toral" guides readers toward "ways to not only reformulate our relationships to language but, through language, to revise our relationships with the land" (Legler, 23).

2. It seems a bit insulting for this great American environmentalist lyric to reside in *Poetry for Dummies*, and yet there it is, incongruously offset next to a bug-eyed cartoon figure in the margin who points his finger in the air and commands readers to "Read Aloud" Francis's "Silent Poem," a composition that John Timpane and Maureen Watts, editors for Hungry Minds, Inc., admit "is almost impossible to read silently." Still, Tim-pane and Watts do produce commentary that leans toward an environmentalist reading: "It turns into a striking, descriptive piece, with a sense of time passing. . . . The short pauses between the words help you—perhaps even gently *force* you—to visualize each hard, concrete thing as you come to it, until, as the reader, you are in the midst of na-ture" (Timpane and Watts, 26).

3. Alberta Turner suggests that the strategically grouped words and stanzas consti-tute an intentional narrative, a "statement of human futility," that contrasts the "easy grace and sure survival of natural species . . . and the man's own hard, marginal sub-sistence." Turner adds, "This man takes his very presence and permanent place in the *nature* of *things*—upon the bedrock. He has *made* his place. The woodchuck, firefly and jewelweed have merely *occurred* in theirs. . . . Craft alone could not have kept this poem from becoming a Currier and Ives calendar towel if a particularly honest percep-tion had not gathered wood and weed, pitchfork and thunder, man and rock and clanged

them together" (Turner, 318). Karen F. Stein notes, "Alliteration and assonance knit this poem together in a patchwork pattern of sounds, and the traditional New England images generate a visual pattern, . . . a picture of the New England rural landscape" (Stein, 221). Unmoved by Francis's stylistics of silence—Alan Sullivan claims that Francis, late in life, "tried to adapt to poetic fashion" with "poems consisting entirely of nouns—a waste of his gift" (Sullivan, par. 20).

4. In terms of the *res* and *verba* divide, David Gilcrest's discussion of Leonard Scigaj's definition of the ecopoet is helpful in reading a poem such as "Silent Poem." Scigaj argues that the ecopoet directs our attention "beyond the printed page toward firsthand experiences that approximate the poet's intense involvement in the authentic experience that lies behind his originary language.'" "Such a gesture," Gilcrest writes, "is predicated on experience of the world unmediated by language." Scigaj adds, "Human language is much more limited than the ecological processes of nature . . . . Ecopoets recognize the limits of language while referring us in an epiphanic moment to our interdependency and relatedness to the richer planet whose operations created and sustain us." Gilcrest concludes, "By quieting the mind, silencing the chatter of language, repudiating its propensity for attachment and discrimination, one experiences loss of self and a concomitant ecstatic synthesis in the world" (Gilcrest, "Regarding Silence" 18–19).

5. Writing of Francis's poetry generally, Richard Wilbur commented, "The reader has scarcely any sense of a poet standing between him and the scene, brandishing a rhetoric and offering clear interpretations. Because the poet thus effaces himself, because he writes so transparently, his formal felicities—though they have their effect—are not felt as part of a performance" (Wilbur, "On Robert," 320). Four years after the publication of "Silent Poem" in book form, Dolores Whitney, in her thesis, "Robert Francis: Best Neglected Poet," observed, "Francis rarely uses the word 'I' in a poem. His intention was to let the words speak for themselves—to *be* the poem, with no comment on them from the poet." Writing during a period historically associated with postmodern experimentation, Whitney identified Francis's art as an "entirely new field of poetry" that had "yet to be explored" (Whitney, 45, 59, 81).

CHAPTER 8. VALHALLA

1. Greg Garrard traces the environmental apocalyptic tradition to early sources. "It seems likely that the distinctive construction of apocalyptic narrative that inflects much environmentalism today began around 1200 BCE," Garrard writes, "in the thought of the Iranian prophet Zoroaster, or Zarathustra" (Garrard, 88). Lawrence Buell adds, "The concept of annihilative apocalypse is as old as Lucretius," and, in terms of the American tradition of literary apocalypticism, Buell lists several examples: Carson's *Silent Spring*; Silko's *Ceremony*; Cooper's *Crater*; Poe's *Eureka*; Melville's *Moby-Dick*; Twain's *Connecticut Yankee*; Donnelly's *Caesar's Column*; London's *Iron Heel*; Faulkner's *Absalom! Absalom!*; and West's *Day of the Locust* (Buell, *Environmental Imagination*, 299, 301). As recently as 2006, Robin Collin has explored the "apocalyptic vision" that "currently influences contemporary US government law and policy" but which "isn't new or unique to the US or even Christianity" (Collin, 1). To Collin, American environmental apocalypticism remains primarily a literary enterprise: "The truly influential role to be played in this transformation is in the hands of artists, writers, and imaginative people, not lawyers or politicians whose work nearly always follows and rarely leads" (Collin, 2–3).

2. In a 1938 review, David Morton lauded *Valhalla* as a "forthright and realistic narrative that moves in an atmosphere of lyrical and ideal beauty and intensity" (Morton, 10). In 1977, Paul Ramsay crowned it "one of the best long poems of the century"

(Ramsey, 538). Not quite as moved in the twenty-first century, Alan Sullivan reflects on *Valhalla* as a "blank-verse melodrama" that "runs more than seventy pages" and "dominates" Francis's second collection (Sullivan, par. 13).

3. Matthew Cooperman cites Buell, who appears to describe young Leif's dilemma. Buell observes, "If, as environmental philosophers contend, western metaphysics and ethics need revision before we can address today's environmental problems, then environmental crisis involves a crisis of the imagination the amelioration of which depends on finding better ways of imaging nature and humanity's relations to it" (Cooperman, 209).

4. The concept of opposing cultural and environmental dualistic thinking abounds in both creative and theoretical ecological literature. Christopher Manes has pointed out that "many primal groups have no word for wilderness and do not make a clear distinction between wild and domesticated life, since the tension between nature and culture never becomes acute enough to raise the problem" (Manes, 18). Bruce Allen, in his work on Ando Shoeki, eighteenth-century Japanese philosopher from the Edo Period, observes that Shoeki, "rejected the hierarchical principles of Confucianism" and "anthropocentric views of the world, replacing them with a new view of harmonious relations between plants, animals, and the earth itself." Allen quotes Shoeki: "Direct cultivation . . . is the kernel of the true way of Nature. On the contrary it is concealed behind dualistic theories about Heaven and Earth, sun and moon, man and woman, lord and subject, Buddha and masses, high and low, noble and mean, good and bad, the fallacy of which lies in failing to see that each of these pairs is not two but one. The True Way is the operation of Nature, and the Way is but One. Heaven and earth constitute a single body, the sun and the moon one God, five kinds of cereal one grain, man and woman one character . . . life and death one way, pleasure and pain one feeling, and joy and anger one sense" (Allen, 306).

5. Across centuries and cultures (including the current one, in which e-mail, the Internet, cell phones, and text-messaging challenge the utility of paper and printed language while chipping away at human levels of socioenvironmental sensitivity), writers revisit the impact of language on human intelligence and non-human preservationism. The concluding narrative concerning Thoth and Ammon in Plato's *Phaedrus* provides a classic example. Bruce Allen notes that Ando Shoeki, in his writings, claimed that "the development of writing" triggered a cultural shift when humans began "to think of themselves as separate from the language and body of the natural world." Further, Allen writes that Shoeki "lament[ed]" the "invention of writing in China," which led to the "state of peace and war." To Shoeki, it was the "complicated Chinese ideographic characters" that led to a "fetishism" that caused "people to lose their ability to hear and respond to the sounds of the nonhuman world" (Allen, 309–310).

6. An entry in Francis's journal (May 21, 1934) describes a "rare experience" that appears to have provided him with the real-life templates for some of the characters and situations in *Valhalla*. In it, Francis writes of climbing "Butter Hill" and walking to the "Gulf Road" to visit a family he knew there, his motivation being that he was "curious" to see the independent "printing establishment" owned and run by a young man named "Allen Ross." On arriving, Francis met not Allen, but his father, a "sturdy, sunburned, spirited man." Francis writes of being impressed with the "praiseworthy as well as picturesque" small press operation in the farmhouse attic, but more so by Mr. Ross, a father of nine children. "I don't know when a man has made a better impression on me than the father," Francis notes. Among other things, Mr. Ross proudly told Francis that he was happy to "spend his money in food for his family rather than in doctor's bills." At Francis's departure, Mr. Ross pointed to "his apple trees still in bloom" and said, "I'm as much interested in those as Allen is in his printing" (*Travelling in Amherst*, 34–35). The apple trees, the young printing entrepreneur, the value of printing versus nature argument, the independent father figure—all inform *Valhalla*.

7. In order to understand the impact of the ending of *Valhalla*, we might refer to Daniel Spencer's focus on Donna Haraway's definition of "amodernity." Haraway argues that an "*a*modern view" of culture "refuses beginnings (universalizing origin narratives that hide the social history that produced them), enlightenments, and endings (especially apocalyptic scenarios about the end-times from which we derive our current identity and meaning." As if describing the structure of *Valhalla*, she adds, "The world has always been in the middle of things, in unruly and practical conversation, full of action and structure by a startling array of actants and of networking and unequal collections. . . . The shape of amodern history will have a different geometry, not of progress, but of permanent and multi-patterned interaction through which lives and worlds get built, human and unhuman. . . . Apocalyptic narratives that predict the return of 'the sacred image of the same' are inadequate because the system is not closed and the world is not full. It's not a 'happy ending' we need, but a non-ending." (Spencer 94–95).

8. John Gatta puts the American apocalyptic narrative in similarly religious terms, focusing on the "religious tradition" of the "American jeremiad," a "warning that sinful violations of the earth may be revealed in future cataclysm" (Gatta, 146).

9. Revelation 5:12 describes creation's apocalyptic end in terms of cosmic, atmospheric, and geospheric upheaval: "And I beheld when he had opened the sixth seal, and, lo, there was a great earthquake; and the sun became black as sackcloth of hair, and the moon became as blood."

10. John Tallmadge reminds us of the prominent place of non-human actors in environmental literature: "The salient feature of environmental literature is that nature is not merely a setting or backdrop for human action, but an actual factor in the plot, that is, a character and sometimes even a protagonist" (Tallmadge, 282).

11. Robin Collin's general discussion of the American apocalyptic vision corresponds with the impact of Francis's final scene: "Others considering the meaning and function of these visions in human psychology conclude that the apocalypse is not so much about death or annihilation as it is about the perceived need for change on a massive scale. These visions occur when we come to the end of a way of life that has become impossible for some reason. . . . The Apocalypse seems an understandable response to an explosive desire to unshackle ourselves from systems which have become corrupt and dysfunctional. The Apocalyptic vision is a comprehensive, visceral response to that realization" (Collin, 2).

Bibliography

Adams, Raymond. Letter to H. Leland Varley. July 7, 1970. Robert Francis Papers, W. E. B. DuBois Library, University of Massachusetts.

Adelson, Glenn, and John Elder. "Robert Frost's Ecosystem of Meanings in 'Spring Pools.'" *ISLE* 13.2 (2006): 1–17.

Allen, Bruce. "Ando Shoeki, Ecology and Language." In Harrington and Tallmadge, *Reading under the Sign of Nature,* 299–314.

Anonymous. Review of *A Certain Distance,* by Robert Francis. *Small Moon.* 1976. Robert Francis Papers. W. E. B. DuBois Library, University of Massachusetts.

Armbruster, Karla. "Bring Nature Writing Home: Josephine Johnson's *The Inland Island* as Bioregional Narrative." In Harrington and Talmadge, eds., *Reading under the Sign of Nature,* 3–23.

Ashton, Jennifer. *From Modernism to Postmodernism: American Poetry and Theory in the Twentieth Century.* Cambridge: Cambridge University Press, 2006.

Barr, John. "American Poetry in the New Century." *Poetry* 188 (2006): 433–41.

Bate, Jonathan. *The Song of the Earth.* Boston: Harvard University Press, 2002.

Bell, David. "Eroticizing the Rural." In Phillips, Watt, and Shuttleton, *De-Centering Sexualities,* 83–101.

Benét, William Rose. "Other New Books: *Stand with Me Here.*" Review of *Stand with Me Here,* by Robert Francis. *Times Literary Supplement,* January 9, 1937: 30.

Bloom, Harold. "The Anxiety of Influence." In Leitch, *Norton Anthology of Theory and Criticism,* 1794–1805.

Bly, Robert. "The Grace of Indirection." *Painted Bride Quarterly* 35 (1988): 101–5.

Booth, Philip. Review of *The Orb Weaver,* by Robert Francis. *Christian Science Monitor* (February 21, 1960): 11.

Boston Post. Review of *We Fly Away,* by Robert Francis. (September 26, 1948.) Robert Francis Papers. W. E. B. DuBois Library. University of Massachusetts.

Branch, Michael P. "The V. E. C. T. O. R. L. O. S. S. Project." *Isotope* 5.2 (2007): 4–9.

Branch, Michael P., and Scott Slovic, eds. *The ISLE Reader: Ecocriticism, 199–003.* Athens: University of Georgia Press, 2003.

Bryson, J. Scott. *Ecopoetry: A Critical Introduction.* Salt Lake City: University of Utah Press, 2002.

———. *The West Side of Any Mountain: Place, Space, and Ecopoetry.* Iowa City: University of Iowa Press, 2005.

Buell, Lawrence. *The Environmental Imagination: Thoreau, Nature Writing, and the Formation of American Culture.* Cambridge, MA: Belknap Press of Harvard University Press, 1995.

———. "Frost as a New England Poet." In *Cambridge Companion to Robert Frost,* edited by Robert Faggen, 101–22. Cambridge: Cambridge University Press, 2001.

————. *The Future of Environmental Criticism: Environmental Crisis and Literary Imagination.* Malden: Blackwell, 2005.

Cahalan, James P. "Teaching Hometown Literature: A Pedagogy of Place." *College English* 70 (2008): 249–74.

Ceraolo, Michael. "The Base-Stealer Reprised." *Slow Trains,* vol. 5 no. 2. <http://www.slowtrains.com/vol5issue2/ceraolovol5issue2.html>.

Christensen, Laird. "The Pragmatic Mysticism of Mary Oliver." In Bryson, *Ecopoetry,* 135–52.

Collin, Robin Morris. "The Apocalyptic Vision, Environmentalism, and a Wider Embrace." *ISLE* 13.1 (2006): 1–9.

Cooperman, Matthew. "Charles Olson: Archeologist of Morning, Ecologist of Evening." In Harrington and Tallmadge, *Reading under the Sign of Nature,* 217–35.

Cresswell, Tim. *Place: A Short Introduction.* Malden: Blackwell, 2004.

Curran, Beverly. "In Her Element: Daphne Marlatt, the Lesbian Body, and the Environment." In Bryson, *Ecopoetry,* 195–206.

Dickey, James. Letter to Robert Francis. September 6, 1960. Robert Francis Archives. E. S. Bird Library, Syracuse University, New York.

Dickinson, Emily. *The Poems of Emily Dickinson.* Edited by R. W. Franklin. 3 vols. Cambridge, MA: Belknap Press of Harvard University Press, 1998.

————. *Selected Letters.* Edited by Thomas H. Johnson and Theodora Ward. 3 vols. Cambridge, MA: Belknap Press of Harvard University Press, 1998.

Dunning, Stephen, and Robert Francis. "Poetry as (Disciplined) Play." *English Journal* 52 (1963): 601–9.

Eberwein, Jane Donahue. *Dickinson: Strategies of Limitation.* Amherst: University of Massachusetts Press, 1985.

Feder, Helena. "Ecocriticism, New Historicism, and Romantic Apostrophe." In *The Greening of Literary Scholarship: Literature, Theory, and the Environment,* edited by Steven Rosendale, 39–53. Iowa City: University of Iowa Press, 2002.

F. P. R. Review. of *We Fly Away,* by Robert Francis. *Amherst Record.* Robert Francis Papers, W. E. B. DuBois Library, University of Massachusetts. September 30, 1948.

Francis, Robert. *A Certain Distance.* Woods Hole, MA: Pourboire, 1976.

————. *Collected Poems: 1936–1976.* Amherst: University of Massachusetts Press, 1976.

————. *Come Out into the Sun: Poems New and Selected.* Amherst: University of Massachusetts Press, 1966.

————. "Conversation with a Woodchuck." *Christian Science Monitor* October 1, 1947: 18.

————. "Emily and Robert." Unpublished ms. Henry Lyman Personal Collection. Northampton, MA, 1963.

————. "Emily Dickinson: 1965." Unpublished ms. Henry Lyman Personal Collection. Northampton, MA.

————. "Emily Dickinson: Her Posthumous Drama." Unpublished ms. Henry Lyman Personal Collection. Northampton, MA. 1967.

————. "Emily for Everybody." *The New England Review* 1.4 (1979): 505–11.

————. *The Face against the Glass.* Amherst, MA: Self-published, 1950.

————. *Francis on Poetry.* Unpublished ms. Henry Lyman Personal Collection, Northampton, MA. 1985.

———."Frost as Apple Peeler." *New England Review* 1.1 (1978): 32–39.

———. *Frost: A Time to Talk*. Amherst: University of Massachusetts Press, 1972.

———. "Fort Juniper." *Painted Bride Quarterly* 35.1 (1988): 14.

———. *Gusto, Thy Name Was Mrs. Hopkins: A Prose Rhapsody*. Toronto: Chartres, 1988.

———. *Late Fire, Late Snow*. Amherst: University of Massachusetts Press, 1992.

———. Letter to Edward J. Rutmayer. December 7, 1976. Robert Francis Papers. W. E. B. DuBois Library, University of Massachusetts.

———. Letter to George E. McPherson. January 1, 1932. Robert Francis Archives. E. S. Bird Library, Syracuse University, New York.

———. Letter to Jac L. Tharpe. September 15, 1976. Robert Francis Papers. Jones Library, Amherst, MA.

———. Letter to Michael Hamburger. July 19, 1966. Robert Francis Papers. E. S. Bird Library, Syracuse University, New York.

———. Letter to Samuel French Morse. October 19, 1963. Robert Francis Papers. E. S. Bird Library, Syracuse University, New York.

———. *Like Ghosts of Eagles*. Amherst: University of Massachusetts Press, 1974.

———. "Market Hill Road." *The Christian Science Monitor,* March 31, 1951: 8.

———. "The Moon as Entertainment." December 10, 1953. Robert Francis Papers. E. S. Bird Library, Syracuse University, New York.

———. "Mountain Day." October 19, 1953. Robert Francis Papers. E. S. Bird Library, Syracuse University, New York.

———. "Nine Great Ones." August 15, 1948. Robert Francis Papers. E. S. Bird Library, Syracuse University, New York.

———. *The Orb Weaver*. Middleton, CT: Wesleyan University Press, 1960.

———. "Outdoors Indoors." December 23, 1941. Robert Francis Papers. E. S. Bird Library, Syracuse University, New York.

———. *Poems New and Selected*. Amherst, MA: University of Massachusetts Press, 1966.

———. "The Poet." *Christian Science Monitor,* July 29, 1985. February 8, 2008. <http://www.csmonitor.com/1985/0729/ufran.html>.

———. *Pot Shots at Poetry*. Ann Arbor: University of Michigan Press, 1980.

———. *Rome without Camera*. Amherst, MA: Jones Library, 1958.

———. *The Satirical Rogue on All Fronts*. Amherst: University of Massachusetts Press, 1968; repr. Amherst, MA: Ardsley Press, 1984.

———. "Season Between." Robert Francis Papers. E. S. Bird Library, Syracuse University, New York., November 25, 1950.

———.*Stand with Me Here*. New York: Macmillan, 1936.

———. "The Stones in My Life." *Christian Science Monitor,* January 18, 1950: 8.

———. *The Sound I Listened For*. Amherst, MA: Self-published. 1944.

———. "Television with No Antenna." Robert Francis Papers. E. S. Bird Library, Syracuse University, New York, May 6, 1953.

———. *Traveling in Concord*. Unpublished ms. Robert Francis Papers. Henry Lyman- Personal Collection, Northampton, MA.

———. *Travelling in Amherst: A Poet's Journal 1930–1950*. Boston: Rowan Tree, 1986.

————. *The Trouble with Francis*. Amherst: University of Massachusetts Press, 1971.

————. *The Trouble with God*. West Hatfield, MA: Pennyroyal Press, 1984.

————. *Valhalla and Other Poems*. New York: Macmillan, 1938.

————. "A Walk Between Two Days." Robert Francis Papers. E. S. Bird Library, Syracuse University, New York, November 17, 1949.

————. *We Fly Away*. New York: Swallow, 1948.

Garrard, Greg. *Ecocriticism*. New York: Routledge, 2004.

Gatta, John. *Making Nature Sacred: Literature, Religion, and Environment in America from the Puritans to the Present*. Oxford: Oxford University Press, 2004.

Gerhardt, Christine. " 'Often seen—but seldom felt': Emily Dickinson's Reluctant Ecology of Place." *Emily Dickinson Journal* 15.1 (2006): 56–78.

Gifford, Terry. "Gary Snyder and the Post-pastoral." In Bryson, *Ecopoetry*, 76–87.

Gilbert, Sandra M., and Susan Gubar. "Infection in the Sentence: The Woman Writer and the Anxiety of Authorship." In Leitch, *Norton Anthology of Theory and Criticism*, 2023–25.

Gilcrest, David. "Regarding Silence: Cross-cultural Roots of Ecopoetic Meditation." In Bryson, *Ecopoetry*, 18–29.

Gillman, Richard. "The Man Robert Frost Called 'The Best Neglected Poet.'" Robert Francis Papers, W. E. B. DuBois Library, University of Massachusetts., March 10, 1985.

————. "Amherst Poet's First Novel Is Homely Tale." Review of *We Fly Away*, by Robert Francis. *Daily Hampshire Gazette*. September 10, 1948. Robert Francis Papers. W. E. B. DuBois Library, University of Massachusetts.

————. Introduction. *Travelling in Amherst: A Poet's Journal 1930–1950*. By Robert Francis. Boston: Rowan Tree, 1986, vii–xviii.

Glotfelty, Cheryll, and Erich Fromm, eds. *The Ecocriticism Reader*. Athens: University of Georgia Press, 1996.

Graham, David. "Millimeters Not Miles: The Excellence of Robert Francis." *Painted Bride Quarterly* 35.1 (1988): 79–93.

Hall, Donald. Letter to Robert Francis. July 7, 1977. Robert Francis Archives. E. S. Bird Library, Syracuse, University, New York.

————. "On 'His Running My Running.'" In Walker, *Poets Reading*, 311–12.

————. "Two Poets Named Robert." *Ohio Review* 18.3 (1977): 110–25.

Hamburger, Michael. *Art as Second Nature: Occasional Pieces 1950–74*. Manchester, UK: Carcanet New Press, 1975, 150–52.

Haralson, Eric. "Introduction." In *Reading the Middle Generation Anew: Culture, Community, and Form in Twentieth-century American Poetry*, edited by Eric Haralson. Iowa City: University of Iowa Press, 2006.

Harrington, Henry, and John Tallmadge, eds. *Reading under the Sign of Nature: New Essays in Ecocriticism*. Salt Lake City: University of Utah Press, 2000.

Hart, George. "Postmodernist Nature/Poetry: The Example of of Larry Eigner." In Harrington and Tallmadge, *Reading under the Sign of Nature*, 315–22.

Heise, Ursula K. "The Hitchhiker's Guide to Ecocriticism." *PMLA* 121 (2006): 503–16.

Hoffman, Frederick. *The Modern American Novel in America: 1900–1950*. Whitefish, MT: Kessinger Publishing, 2009. First published 1951 by Henry Regnery Publishing, 1951.

Howes, Victor. "Of Bulldozers and Bees: A Celebration in Poetry." *Christian Science Monitor*, June 14, 1974, F4.

Ingebresten, Edward J. "Biographers of Robert Frost." In *The Robert Frost Encyclopedia*, edited by Nancy Lewis Tuten and John Zubizarrata, 25–31. Westport, CT: Greenwood Press, 2001.

Kelly, David. "The Base Stealer: Robert Francis, 1948." In *Poetry for Students*, edited by Jennifer Smith and Elizabeth Thomason, vol. 12, 29–42. New York: Gale, 2001.

Kikel, Rudy. Review of *A Certain Distance*, by Robert Francis. *The Advocate*. Robert Francis Papers. W. E. B. DuBois Library, University of Massachusetts.

Larsen, Lance. *In All Their Animal Brilliance*. Tampa, FL: University of Tampa Press, 2005.

Legler, Gretchen. "Toward a Postmodern Pastoral: The Erotic Landscape in the Work of Gretel Ehrlich." In Branch and Slovic, *The ISLE Reader*, 22–32.

Leitch, Vincent B., ed. *The Norton Anthology of Theory and Criticism*. New York: Norton, 2001.

Leopold, Aldo. "From a *Sand County Almanac*." In *The Norton Book of Nature Writing*, edited by Robert Finch and John Elder, 401–21. New York: Norton, 1990.

———. *A Sand County Almanac*. New York: Oxford University Press, 1966.

Lizak, Leonard. "Robert Francis: A Trinity of Values, Nature, Leisure, Solitude." MEd. thesis. California, PA: California University of Pennsylvania, 1966.

Loiseau, Jean. *The Poetry of Robert Francis: A New England Poet*. Diplôme d' Etudes Supérieures. Bordeaux, France, 1962.

Love, Glen A. "Revaluing Nature." In Glotfelty and Fromm, *The Ecocriticism Reader*, 277–34. Athens: University of Georgia Press, 1996.

Lyman, Henry. Personal interview. May 18, 2007.

———. Introduction. In *After Frost: An Anthology of Poetry from New England*, edited by Henry Lyman, 27–29. Amherst: University of Massachusetts Press, 1996.

Manes, Christopher. "Nature and Silence." In Glotfelty and Fromm, *The Ecocriticism Reader*, 15–26.

McCloskey, Mark. "Five Poets." *Poetry* 58 (1966): 272–76.

McGinnis, Michael Vincent, ed. *Bioregionalism*. London: Routledge, 1999.

———. "Boundary Creatures and Bounded Spaces." In McGinnis, *Bioregionalism*, 71–78.

———. "A Rehearsal to Bioregionalism." In McGinnis, *Bioregionalism*, 3–11.

McLennan, Gordon. "Afterword." In *Gusto, Thy Name Is Mrs. Hopkins: A Prose Rhapsody*, by Robert Francis. Toronto: Chartres, 1988, 41–49.

McNair, Wesley. "The Triumph of Robert Francis." In *Mapping the Heart: Reflections on Place and Poetry*. Pittsburgh: Carnegie Mellon University Press, 2003.

Meeker, Joseph. *The Comedy of Survival*. New York: Scribner's, 1974.

Merrill, James. Letter to Robert Francis. March 24, 1961. Robert Francis Papers. E.S. Bird Library, Syracuse University, New York.

Moore, Marianne. Letter to Leone Barron. December 31, 1966. Robert Francis Archives. W. E. B. DuBois Library, University of Massachusetts.

———. Letter to Robert Francis. May 11, 1956. Robert Francis Archives. W. E. B. DuBois Library, University of Massachusetts.

————. Letter to Wesleyan University Press. November 10, 1960. Robert Francis Archives. W. E. B. DuBois Library, University of Massachusetts.

————. *The Selected Letters of Marianne Moore.* Edited by Bonnie Costello. New York: Knopf, 1997.

Morris, Jeff. "Reviews: *A Certain Distance.*" *Open Places* 27 (1979): 59.

Morton, David. Review of *Valhalla and Other Poems. New York Times Book Review.* 23 October 23, 1938: 10.

Nelson, Howard. "Moving Unnoticed: Notes on Robert Francis's Poetry." In *Hollins Critic* 14.4 (1977): 1–12.

Oelschlager, Max. *The Idea of Wilderness.* New Haven: Yale University Press, 1991.

Oliver, Mary. *New and Selected Poems.* Vol. 2. Boston: Beacon, 2005.

O. O. Review of *We Fly Away*, by Robert Francis. December 19, 1948. Robert Francis Papers. W. E. B. DuBois Library, University of Massachusetts.

Parini, Jay. *Robert Frost: A Life.* New York: Holt, 1999.

Phillips-Cubbage, Elinor. "Robert Francis: A Critical Biography." MA thesis. Eastern Connecticut State University, 1975.

Phillips, Richard, and Diane Watt. Introduction. In Phillips, Watt, and Shuttleton, *De-centering Sexualities,* 1–9.

Phillips, Richard, Diane Watt, and David Shuttleton, eds. *De-centering Sexualities: Politics and Representation Beyond the Metropolis.* Routledge: London, 2000.

Phillips, Robert. "Robert Francis." *American Poet* 11.1 (1998): 8–10.

"A Poet's Voice Rises from the Archives." *NPR.org.* 2007. National Public Radio. April 6, 2007. <http://www.npr.org/templates/story/story.php?storyId=9262071>.

Quetchenbach, Bernard. *Back from the Far Field: American Nature Poetry in the Late Twentieth Century.* Charlottesville: University Press of Virginia, 2000.

Quinn, Fran. Introduction. In *Painted Bride Quarterly* 35 (1988): 5–7.

Ramsay, Paul. "Faith and Form: Some American Poetry of 1976." *Sewanee Review* 85 (1977): 537–38.

Ricca, Brad. "Emily Dickinson: Learn'd Astronomer." *Emily Dickinson Journal* 9.2 (2000): 96–108.

Rich, Adrienne. "Vesuvius at Home. The Power of Emily Dickinson." In *Adrienne Rich's Poetry and Prose,* selected edition, edited by Barbara Charlesworld Gelpi and Albert Gelpi, 177–94. New York: Norton, 1993.

Rizzo, Steven. "Francis's 'Pitcher.'" *Explicator* 65.2 (2007): 111–16.

Rueckert, William. "Literature and Ecology. An Experiment in Ecocriticism." In Glotfelty and Fromm, *The Ecocriticism Reader,* 105–23.

Rugg, Winifred King. Review of *We Fly Away*, by Robert Francis. *Christian Science Monitor* December 22, 1948. Robert Francis Papers. W. E. B. DuBois Library, University of Massachusetts.

Rukeyser, Muriel. *Saturday Review.* October 1, 1966. Robert Francis Papers. W. E. B. DuBois Library, University of Massachusetts.

Salkey, Andrew. Review of *Like Ghosts of Eagles*, by Robert Francis. n.d. Robert Francis Papers. W. E. B. DuBois Library, University of Massachusetts.

Scigaj, Leonard. "Panentheistic Epistemology: The Style of Wendell Berry's *A Timbered Choir.*" In Bryson, *Ecopoetry,* 117–34.

————. *Sustainable Poetry: Four American Ecopoets.* Lexington: University Press of Kentucky, 1999, 41.

Sedgwick, Eve Kosofsky. *The Epistemology of the Closet*. Berkeley: University of California Press, 1991.

Shaw, Robert B. "Robert Francis: 1901–1987." In *American Writers: A Collection of Literary Biographies. Supplement IX: Nelson Algren to David Wagoner*, edited by Jay Parini, 75–92. New York: Scribner's-Gale, 2002.

Shuttleton, David. "The Queer Politics of Gay Pastoral." In Phillips, Watt, and Shuttleton, *Decentering Sexuality*, 125–46.

Sides, Charles R. "Freedom to Fastidious Form": Theory, Form, and Theme in the Poetry of Robert Francis. MA thesis. Millersville, PA: Millersville State College, 1975.

Sinfield, Alan. "The Production of Gay and the Return to Power." In Phillips, Watt, and Shuttleton, *Decentering Sexuality*, 21–36.

Spencer, Daniel T. *Gay and Gaia: Ethics, Ecology, and the Erotic*. Cleveland, OH: Pilgrim, 1996.

Stambuk, Andrew. "Learning to Hover: Robert Frost, Robert Francis, and the Poetry of Detached Engagement." *Twentieth Century Literature* 45 (1999): 534–52.

Stein, Karen F. "Robert Francis 1901–1987." In *Encyclopedia of American Poetry: The Twentieth Century*, edited by Eric L. Haralson, 221–22. Chicago: Fitzroy Dearborn, 2001.

Sullivan, Alan. "Robert Francis: Choices and Compulsions." In *Fresh Bilge: A Salty Journal* January 25, 2005. <http://www.seablogger.com/library/jade/essays.03.htm>.

Tallmadge, John. "Toward a Natural History of Reading." In Branch and Slovic, *The ISLE Reader*, 277–85.

"Terrain.org Interviews Terry Tempest Williams." *Terrain.org*. August 27, 2005. July 22, 2008. <http://www.terrain.org/interview/17/>.

Theroux, Paul. Letter to Robert Francis. February 13, 1967. Robert Francis Papers. W. E. B. DuBois Papers, University of Massachusetts.

Thomashow, Mitchell. "Toward a Cosmopolitan Bioregionalism." In McGinnis, *Bioregionalis*, 121–32.

Thoreau, Henry David. *Walden*. Ticknor & Fields, 1854. Edited by Richard Lenat. June 26, 2006. <http://thoreau.eserver.org/walden1e.html>.

Timpane, John, and Maureen Watts. *Poetry for Dummies*. New York: Hungry Minds, 2001.

Troubetzkoy, Ulrich. "Review of *The Sound I Listened For*," by Robert Francis. *New York Times Book Review*, April 16, 1944, 26.

Tuan, Yi-Fu. "Discrepancies between Environmental Attitude and Behaviour: Examples from Europe and China." In *Ecology and Religion in History*, edited by David Spring and Eileen Spring, 85–104. New York: Harper & Row, 1974.

———. *Topophilia*. New York: Columbia University Press, 1990.

Turner, Alberta. "Permitting Craft." In *Poets Reading*, edited by David Walker, 316–18. Oberlin, OH: Oberlin College Press, 1999.

Varley, H. L. *We Fly Away*: A Novel by Robert Francis. Amherst Author Writes of N. E. College Town." Review of *We Fly Away*, by Robert Francis. Robert Francis Papers. W. E. B. DuBois Papers. University of Massachusetts.

Walker, David, ed. *Poets Reading: The FIELD Symposia*. Oberlin. OH: Oberlin College Press, 1999.

———. "Francis Reading and Reading Francis." In Walker, *Poets Reading*, 28–30.

Wallace, Robert. "The Excellence of 'Excellence.'" *Field* 25.1 (1981): 15–18.

Wang, Hui-Ming. "A Remembrance: Notes on Illustrations." *Painted Bride Quarterly* 35 (1988): 108–10.

Warren, James Perrin. "Contexts for Reading 'Song of the Redwood Tree.'" In Harrington and Tallmadge, *Reading under the Sign of Nature*, 166–73.

White, Lynn. "The Historical Roots of Our Ecologic Crisis." In Glotfelty and Fromm, *The Ecocriticism Reader*, 4–14.

Wilbur, Richard. Introduction. In *Butter Hill.*, by Robert Francis. Amherst, MA: Ardsley, 1984.

———. "On Robert Francis' 'Sheep.'" In Walker, *Poets Reading*, 319–20.

Williams, Terry Tempest. "Terrain.org. Interview, Terry Tempest Williams. <http://www.terrain.org/interview/17>.

Young, David. "Robert Francis (1901–1987)." *The Longman Anthology of Contemporary American Poetry: 1950 to the Present.* 2nd ed., edited by Stuart Friebert and David Young, 62–64. New York: Longman, 1989.

———. "Robert Francis and the Blue Jay." In Walker, *Poets Reading*, 308–11.

Young, Mary. "On the Road." *Valley Advocate* January 9, 1980, 24.

Index